AF800278

Two-Dimensional Metal Halide Perovskites

John S. Colton · Kameron R. Hansen
Editors

Two-Dimensional Metal Halide Perovskites

A Machine-Generated Literature Overview

Springer

Editors
John S. Colton
Department of Physics, Eyring Science Center
Brigham Young University
Provo, UT, USA

Kameron R. Hansen
Department of Chemistry
University of Utah
Salt Lake City, UT, USA

ZwiCoat Materials Innovations
Iowa City, IA, USA

ISBN 978-981-99-7829-8 ISBN 978-981-99-7830-4 (eBook)
https://doi.org/10.1007/978-981-99-7830-4

© The Editor(s) (if applicable) and The Author(s), under exclusive license to Springer Nature Singapore Pte Ltd. 2024

This work is subject to copyright. All rights are solely and exclusively licensed by the Publisher, whether the whole or part of the material is concerned, specifically the rights of translation, reprinting, reuse of illustrations, recitation, broadcasting, reproduction on microfilms or in any other physical way, and transmission or information storage and retrieval, electronic adaptation, computer software, or by similar or dissimilar methodology now known or hereafter developed.
The use of general descriptive names, registered names, trademarks, service marks, etc. in this publication does not imply, even in the absence of a specific statement, that such names are exempt from the relevant protective laws and regulations and therefore free for general use.
The publisher, the authors and the editors are safe to assume that the advice and information in this book are believed to be true and accurate at the date of publication. Neither the publisher nor the authors or the editors give a warranty, expressed or implied, with respect to the material contained herein or for any errors or omissions that may have been made. The publisher remains neutral with regard to jurisdictional claims in published maps and institutional affiliations.

This Springer imprint is published by the registered company Springer Nature Singapore Pte Ltd.
The registered company address is: 152 Beach Road, #21-01/04 Gateway East, Singapore 189721, Singapore

If disposing of this product, please recycle the paper.

Preface

Trends within the scientific community ebb and flow with currents just as strong as those in fashion and the arts. The rise in popularity of 2D metal halide perovskites (MHPs) over the last decade can be squarely attributed to the confluence of two of this decade's most aggressively researched material systems, 2D semiconductors and 3D metal halide perovskites.

Research efforts on the former first escalated in 2010 when bulk MoS_2 was exfoliated down to a single monolayer, causing the material to transition from an indirect to direct bandgap semiconductor. Just two years later in 2012, metal halide perovskites experienced their "viral moment" when a single-junction perovskite solar cell was shown to be 10.9% efficient. The subsequent 2010s and early 2020s have seen tens of thousands of articles on these two subjects and great breakthroughs have been achieved. Naturally, 2D MHPs, having one foot in the 2D world and the other in the perovskite world, also witnessed a great rise in research attention over this time period.

The crystal structures referred to as 2D MHPs, those being "Ruddlesden-Popper" and "Dion-Jacobson", were first reported in 1958 and 1981, respectively. Some years later in the 1990s, it was discovered that the incorporation of metal atoms and halides into these structures produced crystals with impressive electronic, absorption, and emission properties, and moderate research efforts followed. However, it wasn't until the 2010s when interest in these materials saw a meteoric rise, as shown in Fig. 1. By way of housekeeping, we note that what we and many others refer to as MHPs also go by other names in the literature, such as halide perovskites (HPs), hybrid organic–inorganic perovskites (HOIPs), organic–inorganic hybrid perovskites (OIHPs), or metal halide semiconductors (MHSs).

Two-dimensional perovskites are uniquely impressive in their own respect. To date, they remain the only known multiple quantum well material that crystalizes naturally. They are highly tunable, and are among the most absorptive and emissive semiconductors. Still, whether the community's research efforts pay dividends in the form of transformative technologies remains an open question.

We were approached by Springer Nature with the proposition to be editors for a book created by a human–machine collaboration. The proposed formula was

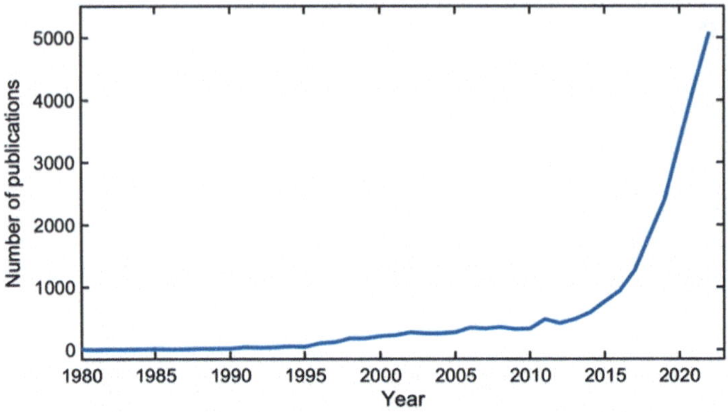

Fig. 1 Number of publications on 2D MHPs from 1980 to 2022. These data were obtained by summing search results for "Dion-Jacobson", "Ruddlesden-Popper" and "2D perovskite"

simple—as editors, we decide the chapter topics and provide the most relevant articles to be featured. Next, the machine generates article summaries and we write chapter introductions. With the dramatic uptick of interest in 2D MHPs, we felt that the field could benefit from such a book on this topic.

In this human–machine collaborative text, in Chap. 1 we begin by providing a broad overview of two-dimensional metal halide perovskites. The following ones, Chaps. 2 and 3, focus on the optoelectronic and lattice properties, respectively. And finally, with Chaps. 4–7, we feature articles that are centered on applications of 2D MHPs, both realized and hypothetical. We strive to highlight the most important advancements made in Springer Nature journals on 2D MHPs. We hope that the book can serve as an effective introduction for newcomers to this field as well as a helpful resource for established researchers seeking topic summaries.

Provo, UT, USA John S. Colton
Iowa City, IA, USA Kameron R. Hansen

Contents

About the Editors

Dr. John S. Colton completed his undergraduate studies at Brigham Young University, graduating with B.S. degrees in Physics and Mathematics in 1994. He received his Ph.D. in condensed matter physics from U.C. Berkeley in 2000. Later, he was a postdoctoral researcher at the Naval Research Laboratory in Washington, D.C. from 2001 to 2003, studying optically detected and controlled semiconductor spintronics. He joined the faculty at University of Wisconsin-La Crosse from 2003 to 2007, after which he returned again to Brigham Young University, as faculty. John's research throughout his career has involved studying semiconductors through optical methods. Since 2021, he has focused on hybrid organic-inorganic metal halide perovskites. In addition to research, John also enjoys teaching and has won several awards, including receiving a 2024 Outstanding Teaching Award from BYU's College of Computational, Mathematical, and Physical Sciences. John was also appointed as a 2017 BYU Alcuin Fellow and selected as the 2006–2007 Wisconsin Teaching Fellow from UW-La Crosse.

Dr. Kameron R. Hansen graduated from Brigham Young University with a B.S. in Physics and a minor in Mathematics, where he studied quantum dots and wrote an undergraduate thesis on synthesis procedures to grow various types of quantum dots inside spherical proteins. Later, he completed his M.S. degree in chemical physics from Columbia University and was awarded an NSF Graduate Research Fellowship. He worked in Dr. Xiaoyang Zhu's laboratory, researching next generation semiconductors such as metal halide perovskites and monolayer transition metal dichalcogenides. Most recently, Kameron earned his Ph.D. in physical chemistry under the supervision of Dr. Luisa Whittaker Brooks (Utah, Chemistry) while closely collaborating with Dr. John S. Colton (BYU, Physics). His doctoral and postdoctoral research has been multidisciplinary, utilizing both experimental and computational approaches to study two-dimensional perovskites. Kameron currently works as the President and CTO of ZwiCoat Materials Innovations, an Iowa City-based startup developing anti-fouling coatings for medical implants.

Chapter 1
An Introduction to 2D Metal Halide Perovskites

Introduction by Editor

2D metal halide perovskites are neither fully two-dimensional nor fully perovskites. Strictly-speaking, the term "perovskite" refers to the ABX_3 crystal structure (such as $MAPbI_3$ in Fig. 1.1). The B site is typically a metal atom and the X site either an oxygen or a halogen atom. In metal halide perovskites, in which the B site is a metal and the X site a halogen, the A site is typically a large metal atom such as cesium or a small organic molecule such as methylammonium (MA). As larger organic cations such as the butylammonium (BA) featured in Fig. 1.1 are incorporated into the structure in an A' site, the metal halide octahedral units separate into layers to either form a Ruddlesden-Popper structure ($A'_2A_{n-1}B_nX_{3n+1}$) or a Dion-Jacobson structure ($A'A_{n-1}B_nX_{3n+1}$), where n denotes the number of metal halide layers between the A'-site cations. These are referred to as 2D or quasi-2D MHPs.

The A'-site molecules are electrically insulating with bandgaps of ~ 4–7 eV while the metal halide inorganic framework is semiconducting with bandgaps of ~ 2–4 eV. As such, 2D MHPs are often classified as multiple quantum well structures, or equivalently, semiconductor-insulator superlattices. Of course, confinement effects are amplified as the width of the well layer decreases to the lower limit of $n = 1$.

The electronic structure of 2D MHPs can be crudely approximated as that of the metal halide layer since the semiconducting BX_4 sheets typically have negligible electronic coupling to the molecular orbitals in the organic layers. The valence bands are largely comprised of the X-site halide's highest-occupied p orbital anti-bonded to the B-site metal's highest-occupied s orbital. The conduction band, on the other hand, primarily consists of the B-site metal's highest-occupied p orbital. The A'-site molecules that comprise the barrier layer have a comparatively larger bandgap energy and lower dielectric constant. Similar to inorganic GaAs-based quantum wells, the mismatch between the valence and conduction bands at the barrier/well interface

© The Author(s), under exclusive license to Springer Nature Singapore Pte Ltd. 2024
J. S. Colton and K. R. Hansen (eds.), *Two-Dimensional Metal Halide Perovskites*,
https://doi.org/10.1007/978-981-99-7830-4_1

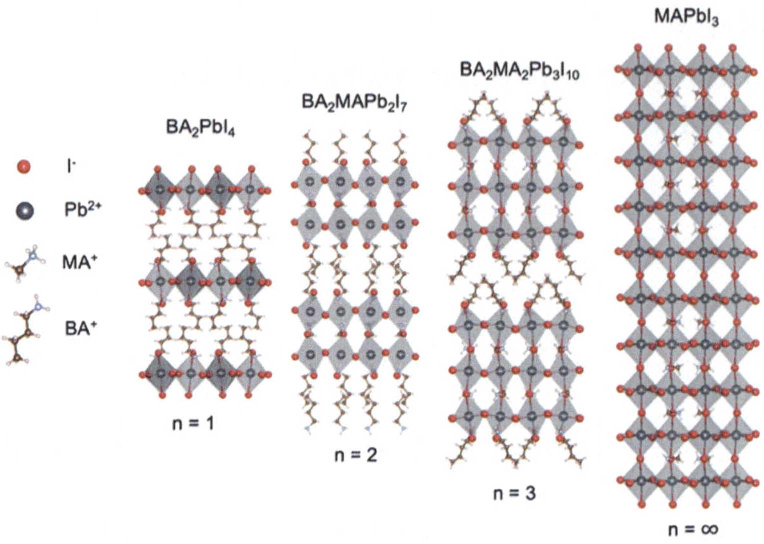

Fig. 1.1 Ruddlesden-Popper crystal structures where $n = 1, 2, 3$, and ∞

creates a quantum confinement effect. This increases the electron and hole wavefunction overlap and strengthens Coulomb interactions. Excitons in 2D MHPs are unique, however, even in comparison to other quantum well systems in that they experience both strong lattice interactions and an image-charge effect from the mismatch in dielectric environments between the organic and inorganic layers. The sum of these effects results in a tightly-bound exciton-polaron (a "small polaron") with an exciton binding energy typically around 200 meV.

Remarkably, unlike other classes of multiple quantum wells (e.g. CdTe-based or GaAs-based), 2D MHPs naturally crystallize into their alternating low bandgap/high bandgap superstructure. More remarkably still, nature offers a large catalog of molecules and atoms that assemble into the Ruddlesden-Popper and Dion-Jacobson 2D frameworks. For example, over 50 different molecules have successfully been incorporated into the A'-site. Lead, tin, and germanium commonly occupy the B-site metal atom, while the X-site halide atom can vary between iodine, bromine, and chlorine. As such, 2D MHPs are highly tunable and this is one of their major advantages. The influence of these compositional degrees of freedom on the optoelectronic properties has occupied a great deal of the research efforts on 2D MHPs over the last decade.

Crystallization strategies for 2D MHPs span a wide gamut. Most commonly, thin films consisting of 2D MHP crystallites are made by dissolving the constituent precursors (PbI_2, butylammonium, etc) into a solvent, spin-coating onto a substrate and then annealing the substrate. However, thin films can also be fabricated with blade coating or chemical vapor depositions. Creating single crystals of 2D MHPs is also relatively simple. One prevalent method involves gently heating the aforementioned precursor solution until the solvent evaporates, leading to the formation of a 2D MHP crystal. Another approach entails dissolving the precursors in an acid such as HI, HBr, or HCl, dissolving the B-site metal atoms and A-site organic cations at a near-boiling temperature, and then allowing the solution to cool slowly. With the excess of halide atoms, millimeter- to centimeter-sized single crystals will precipitate out of the acidic solution. The ease of processing these materials using solutions near room temperature underscores a significant benefit of MHPs.

We begin this chapter highlighting a review article on the semiconductor physics of 2D MHPs, and the subsequent papers are presented in chronological order. Many of the articles have somewhat narrow, specific aims, e.g. to control the orientation of the perovskite layers relative to the substrate, controlling the *n*-number, etc. Together, the papers serve as an introduction to the synthesis, structure, and characterization of 2D MHPs.

Machine generated summaries

Disclaimer: The auto-summaries have been generated by a recursive clustering algorithm via the Dimensions Auto-summarizer by Digital Science handled by Subject Matter Experts and the external editor. The editor of this book selected which SN content should be auto-summarized and decided its order of appearance. Please be aware that the auto-summaries consist of original sentences, but are not representative of its original paper, since we do not show the full length of the publication. Please note that only published SN content is represented here, and that machine-generated books are still at an experimental stage.

Machine generated keywords: heterostructure, vertical, etc, review, monolayer, stack, twodimensional halide, growth, incorporation, various, physics, novel, physical, atomically, space

Semiconductor Physics of Organic–Inorganic 2D Halide Perovskites

DOI: https://doi.org/10.1038/s41565-020-00811-1

Originators: Blancon, JC., Even, J., Stoumpos, C.C. *et al.* Semiconductor physics of organic–inorganic 2D halide perovskites. *Nat. Nanotechnol.* 15, 969–985 (2020). https://doi.org/10.1038/s41565-020-00811-1

Abstract-Summary

Achieving technologically relevant performance and stability for optoelectronics, energy conversion, photonics, spintronics and quantum devices requires creating atomically precise materials with tailored homo- and hetero-interfaces, which can form functional hierarchical assemblies.

Organic–inorganic two-dimensional halide perovskites (2DPKs) are organic and inorganic two-dimensional layers, which self-assemble in solution to form highly ordered periodic stacks.

In this Review, we discuss the current understanding in the structure and physical properties of 2DPKs from the monolayers to assemblies, and present a comprehensive comparison with conventional semiconductors, thereby providing a broad understanding of low-dimensional semiconductors that feature complex organic–inorganic hetero-interfaces.

Main

State-of-the-art electronic, optoelectronic and photonic technologies based on classical inorganic semiconductors such as silicon and III–V materials and the ability to engineer quantum confined heterostructures have been enabled by fundamental knowledge of their structure and physical properties.

The ability to tailor the 2DPKs structural and physical properties through chemical composition, added to the capacity of 2DPKs to absorb and emit light efficiently, and to carry relatively large charge current in a polycrystalline thin film, are unique advantages, which are usually not associated with solution-processed semiconductors.

We also provide a perspective on the future of 2DPKs applications, addressing key scientific questions in their synthesis, structure and properties that need to be tackled in order to access and exploit the intrinsic physics of these materials, which will enable devices with performances above and beyond the state-of-the-art materials and also pave the path for creating new functionalities in devices.

Structural and mechanical properties

In 2DPKs with $n > 1$ there is still limited literature concerning the tuning of both the organic spacer characteristics (for example, polarizability, valency, rigidity, chirality) and the perovskite layer structure (for example, composition, edge or face sharing octahedra, lattice distortion).

In RP 2DPKs with $n = 1$, the longer the alkyl chain in linear alkylammonium A' cation, the higher the temperature of the phase transition, which involves increased conformational and dynamic disorder (or melting) of the A' organic cations, as well as an increase in the perovskite layer symmetry [1–3].

The low-temperature phase transition is intimately linked to distortions (lattice parameter change and octahedra tilting) of the perovskite layer, which depends more on the thickness and composition of the perovskite layer (n-value) [3, 4].

The ferroelectric-paraelectric phase transition has been attributed to reduced distortions of the perovskite octahedra and disordering of the dynamic organic cations in both the perovskite layers and organic spacers.

Physical properties

Two theoretical studies claimed that point defects in 2DPKs with relatively low chemical potential are either electronically inactive (for example, metal-halide vacancies) or only induce shallow trap states (for example, organic cation excess or deficiency) that can result in n-doping; both these types of defects are relatively passive during transport in devices [5, 6].

In 2DPKs with n = 1, studies by Kitazawa and others have suggested that the photoemission fine structure depends on the A' organic spacer, and that the Stokes shifted photoemission can be either bound-exciton or triplet states, which results in more than one order of magnitude difference in photoluminescence lifetime (about 1 ns to more than 50 ns, respectively) [7].

Artificial MQW composed of layers of $CsPbBr_3$ separated by 1,3,5-tris(N-phenylbenzimidazol-2-yl)benzene (TPBi) barriers yield smaller enhancement of the exciton binding energy in 3-nm-thick wells as compared to what is typically observed in Pb-based 2DPKs, which is due to weaker dielectric confinement effects in $CsPbBr_3$/TPBi MQWs.

Challenges in materials design for integrated devices

The use of 2DPK materials with improved crystallinity, high phase purity, and with surfaces promoting radiative charge carrier recombination, results in significant improvement of their photoluminescence efficiency and stability, possibly offering a solution to improve the capacity of 2DPK films to sustain high current density [8–10].

Promising strategies to address the stability issue in 2DPK-based LEDs include synthesis of 2DPK films with high crystallinity and phase purity, improving the packing of A' molecules in the spacing layers to reduce the loss of A' cations due to the degradation process, fabrication of heterostructures with TMD atomic slabs as capping layers on 2DPK films, surface passivation of the grains in films using a molecular matrix, ultra-thin interface layers in thin-film devices to prevent chemical reactions between perovskite and electrodes during operation [10–13].

Outlook

The knowledge developed recently for non-classical material systems such as atomically-thin TMDs, organic semiconductors and 3DPKs has proven useful for the initial understanding of the structural and physical properties of 2DPKs.

One of the main reasons for the lack of understanding of these mechanisms in 2DPKs is the difficulty to model the complex interplay between the dynamical structural characteristics and physical properties of these systems.

The broad phase space offered by 2DPKs featuring various organic–inorganic hetero-interfaces and diverse structural and physical properties make it challenging to draw a broad and unifying picture that describes the semiconductor physics for this class of materials.

The ability to tailor the physical properties of 2DPKs over a broad range via their elemental composition and structure or through external stimuli, combined with their capacity to preserve most of the nanoscale physics at the meso scale and above, gives an edge to the 2DPK family for applications as compared to other emerging (nano-) semiconductors.

Dimensional Tailoring of Hybrid Perovskites for Photovoltaics

DOI: https://doi.org/10.1038/s41578-018-0065-0

Originators:Grancini, G., Nazeeruddin, M.K. Dimensional tailoring of hybrid perovskites for photovoltaics. *Nat Rev Mater* 4, 4–22 (2019). https://doi.org/10.1038/s41578-018-0065-0

Abstract-Summary

The commercial application of 3D hybrid perovskites is inhibited by their poor stability.

Relative to 3D hybrid perovskites, low-dimensional — that is, 2D — hybrid perovskites have demonstrated higher moisture stability, offering new approaches to stabilizing perovskite-based photovoltaic devices.

2D hybrid perovskites have versatile structures, enabling the fine-tuning of their optoelectronic properties through compositional engineering.

We review recent achievements in combining 3D and 2D perovskites as an approach to simultaneously boost device efficiency and stability, paving the way for mixed-dimensional perovskite solar cells for commercial applications.

Introduction

The layered structure relaxes the limits on the cation dimensions imposed by the Goldschmidt tolerance factor and, in contrast to 3D perovskites, enables bulky R cations to be accommodated between the inorganic sheets [14–16].

Increasing the number of layers ($n > 1$) and introducing a small organic cation (A) induce the formation of multi-layered, 'quasi-2D' perovskites, which converge into the 3D structure for $n = \infty$ (refs [14, 15, 17]).

The exciton binding energy progressively decreases with increasing n, that is, on going from a pure 2D to a fully 3D perovskite structure.

Flexibility of 2D perovskites can therefore be used to modify the optoelectronic behaviour and to direct the application of devices [14, 16, 17].

We then present an overview of the structure and optoelectronic behaviour of quasi-2D ($n > 1$) perovskites, highlighting the advances in using 2D or quasi-2D perovskites to stabilize devices.

Low-dimensional perovskites

The generation of free carriers results from instantaneous exciton dissociation owing to the lower exciton binding energy in quasi-2D perovskites [18, 19], similar to what is observed for 3D perovskites.

Strong excitonic effects in 2D perovskites can influence the interaction of charge carriers with lattice vibrations (phonons), leading to strong electron–phonon coupling [20–23].

Comparing the energy of the photoinduced absorption band of the excitonic transitions and the energy of the free-carrier absorption band, a giant Rashba splitting energy of 40 ± 5 meV was calculated with a Rashba parameter of 1.6 ± 0.1 eV Å. This observation provides important proof of concept for the possibility of manipulating the 2D structure to modulate the structural symmetry and to thus design applications beyond optoelectronics, for example, in the field of spintronics or solar spintronics, which exploit the enhanced spin-to-charge conversion efficiency.

Perovskite device stability

Despite the growing interest in hybrid perovskite solar cells, their commercialization is hindered by severe stability issues, which limit device lifetimes and make these devices incompatible with PV market requirements [24].

The UV component of white light is detrimental to the performance of perovskite devices owing to the susceptibility of the device substrate, in particular the TiO_2 layer, to UV-induced degradation [25].

The higher stability of the devices with the buffer layer is thought to be due to a blocking effect, with the buffer layer preventing the migration of metals from the perovskite surface to the HTM.

Thermal stress also influences device stability, inducing the diffusion of the top metal contact into the perovskite [26].

Substituting spiro-OMeTAD with 9-(2-ethylhexyl)-N,N,N,N-tetrakis(4-methoxyphenyl)-9H-carbazole-2,7-diamine results in increased stability, with the device retaining 95% of its maximum PCE for 1,500 h under inert atmosphere accelerated testing conditions.

Stable PSCs with 2D perovskites

Different strategies to improve vertical charge transport, such as controlling the crystallization direction of 2D perovskites or using multiple cations in a quasi-2D perovskite, have been proposed to increase charge transport and extraction and to deliver high device efficiency along with improved material stability.

The use of PEI in the quasi-2D perovskites improves charge transport, which is mainly attributed to defect passivation, leading to a PCE of 8.77% (under ambient conditions with a relative humidity of ~50% without encapsulation) [27, 28].

If n becomes too large, the formation of 2D perovskite domains is limited, reducing device stability (measured at 90 °C at 50% relative humidity).

On $MAPbI_3$, PEA_2PbI_4 and intermediate mixed MA^+:PEA^+ 2D perovskites [29], the effective charge-carrier mobility was found to decrease with increasing PEA^+ content, going from 25 $cm^2\ V^{-1}\ s^{-1}$ for $MAPbI_3$ to 1 $cm^2\ V^{-1}\ s^{-1}$ for PEA_2PbI_4, owing to the concomitant increase in charge-carrier confinement and the exciton binding energy with the percentage of PEA^+.

Future perspectives

The enhanced moisture stability of 2D perovskites and the possibility of exploiting this to engineer stable PSCs motivate wide-ranging research in the PV field.

Strategies for out-of-plane film growth are required to achieve well-aligned 2D perovskites and to improve charge-transport properties in these materials.

Controlling the interface alignment within 2D/3D hybrid perovskites is paramount for optimizing charge transport and collection.

Using 2D perovskite layers as barriers against water penetration into PSCs certainly has potential in improving the device stability; however, this must be achieved without compromising the device efficiency.

The study and development of 2D barrier layers can also advance research on 3D perovskites.

It is possible to control the energy of the HOMO and LUMO levels of the organic cations so that they can actively participate in the energy band structure of the 2D perovskite as well as in the charge-transfer processes, ultimately modulating the function of devices.

Molecular Engineering of Organic–Inorganic Hybrid Perovskites Quantum Wells

DOI: https://doi.org/10.1038/s41557-019-0354-2

Originators: Gao, Y., Shi, E., Deng, S. *et al.* Molecular engineering of organic–inorganic hybrid perovskites quantum wells. *Nat. Chem.* 11, 1151–1157 (2019). https://doi.org/10.1038/s41557-019-0354-2

Abstract-Summary

We demonstrate the modulation of the electronic structure—and consequently the optical properties—of organic semiconducting building blocks that are incorporated between the layers of perovskites through a facile solution processing step.

The energy and charge transfers between adjacent organic and inorganic layers are shown to be fast and efficient, owing to the atomically flat interface and ultrasmall interlayer distance of the perovskite materials.

The resulting two-dimensional hybrid perovskites are very stable due to protection given by the bulky hydrophobic organic groups.

Main

Two-dimensional (2D) layered halide perovskites have attracted considerable attention as an alternative to III–V compounds for achieving quantum-well structures due to their solution processability, broad structure and property tunability, and excellent optoelectronic properties [15, 17, 30–45].

Two-dimensional perovskite structures can be understood as atomically thin slabs that are cut from the three-dimensional parent structures along different crystal directions, which are sandwiched by two layers of large organic cations [46, 47].

There were a large number of reports in the 1990s regarding 2D perovskites that incorporate insulating aliphatic ammonium cations with large energy band gaps and poor charge transport properties; however, only sparse efforts have been made to incorporate semiconducting organic moieties to tune the overall electronic properties.

The range of organic cations that have been incorporated into the crystal lattice remains limited and a fundamental grasp about how molecular structures influence the overall morphology and properties of the perovskites is still lacking.

Results and discussion

The new monoammonium-terminated conjugated ligands can be used to grow molecularly thin 2D crystals (down to a single-quantum-well thick) directly on Si/SiO_2 substrates from a ternary co-solvent [39].

The PL spectrum of the previously reported 2D perovskite $(BA)_2PbI_4$ is also shown (solid/dashed grey curves); $(2T)_2PbI_4$ exhibits a green colour, $(4Tm)_2PbI_4$ and $(4TCNm)_2PbI_4$ display no PL and $(BTm)_2PbI_4$ shows a red colour.

The valence band maximum of $(BA)_2PbI_4$ was determined to be around −5.89 eV. We assume a similar value for the PbI_4^{2-} layer in the new 2D crystals with conjugated ligands because of their structural similarity.

We anticipate that this work will lead to further studies on the chemistry and physics of 2D hybrid perovskite single and multiple quantum wells for the preparation of materials with properties suitable for example for light harvesting, light emission and charge transport.

Tunable Exciton Binding Energy in 2D Hybrid Layered Perovskites Through Donor–Acceptor Interactions Within the Organic Layer

DOI: https://doi.org/10.1038/s41557-020-0488-2

Originators: Passarelli, J.V., Mauck, C.M., Winslow, S.W. *et al.* Tunable exciton binding energy in 2D hybrid layered perovskites through donor–acceptor interactions within the organic layer. *Nat. Chem.* 12, 672–682 (2020). https://doi.org/10.1038/s41557-020-0488-2

Abstract-Summary

An important target is the tuning of a material's exciton binding energy—the energy binding an electron–hole pair through the electrostatic Coulomb force—independent of its electronic band gap.

For four out of seven n = 1 perovskites, the incorporation of the electron-accepting TCBQ dopant into the organic sublattice containing the electron-donating naphthalene species enabled the tuning of the materials' 1s exciton binding energy.

The naphthalene–TCBQ electron donor–acceptor interactions increased the electrostatic screening of the exciton, in turn lowering its binding energy relative to the undoped perovskite—by almost 50% in one system.

Main

It has previously been shown that charge transfer interactions between the inorganic and organic sublattices may be targeted in the design of low-dimensional perovskite structures, by assembling two- or one-dimensional lead halide networks surrounded by highly conductive organic cations [48, 49].

These strategies have led to a reduction in E_b as the 1s exciton moves closer in energy to the band gap in the range of 130–150 meV. Yet, layered perovskite assemblies are highly sensitive to small changes in the organic moiety, as substitution at a different position or lengthening of the alkylammonium linker by one carbon can change how the lead halide network is arranged [16, 50–53].

By identifying which compounds can incorporate TCBQ in a donor–acceptor fashion, we show how structural variations in undoped 2DLP organic layer packing ultimately determine the ability of molecular doping to affect E_b, and emphasize the tunability of 2DLP structures to achieve precise control over fundamental optoelectronic properties.

Results

To determine the degree to which TCBQ is incorporated in perovskites 1 and 2 as a function of x molar equivalents relative to PbI_2, we monitored the change in layer spacing (Δd) using PXRD.

Although 1 and 2 only differ by one carbon in the organic cation and both show a monotonic increase in layer spacing Δd with x, it is only in compound 2 that we observe a notable shift in E_{1s} with increasing TCBQ.

TCBQ) at a temperature of ≤ 10 K. Broader absorption linewidths at all temperatures also result from increased amounts of dopant, which is not surprising since TCBQ is not uniformly incorporated within the perovskite crystalline domains and should lead to enhanced inhomogeneous broadening.

In regime 3, perovskites 3 and 4 do not incorporate TCBQ into the organic lattice at all, with no change in PXRD layered spacing nor in the optical properties.

Discussion

Among 2DLPs, $n = 1$ structures have the strongest dielectric confinement effects, as both faces of the inorganic sheet are sandwiched between organic cation layers.

This means that the exciton Coulomb interaction will extend into the organic layer, where it is poorly screened due to the low-dielectric environment of the Nap cations.

The overall effect of the TCBQ dopant then creates a more polarizable environment within the organic layer, which will be more effective at screening the Coulomb interaction between the electron and hole.

By fitting the temperature-dependent absorption spectra to a modified two-dimensional Elliott model, we determined that the electronic band gap is not changed by the introduction of TCBQ, despite broadening caused by increasing inhomogeneous disorder with dopant addition; the donor–acceptor interactions are able to target the organic layer selectively.

The doped donor–acceptor interactions have an effect on the dielectric environment experienced by the exciton in this series of 2DLPs.

Conclusions

We have successfully tuned the reduction of exciton binding energy in four Nap-based 2DLPs, through incremental addition of the small molecule acceptor TCBQ.

This strategy exploits the intercalating ability of two-dimensional perovskites with a novel molecular dopant strategy that leads to targeted optoelectronic changes: by selecting favourable organic cations and dopants that are capable of donor–acceptor complexation of intermediate strength, we are able to change the dielectric environment of the insulating organic layer without perturbing the inorganic sublattice.

From the comparison of TCBQ doping effects across a family of seven perovskites, we determined that organic cation packing plays an important role in determining whether Nap derivatives are able to bind with TCBQ to form a stable donor–acceptor complex.

Methods

All single-crystal samples were prepared using vapour diffusion of DCM into γ-butyrolactone solutions containing a 2:1 molar ratio of organic ammonium iodide salt:lead (ii) iodide.

Once spinning ceased, the films were directly transferred to a hot-plate in the glove box and annealed at the prescribed temperature for 10 min.

Samples for composition analysis by SEM-EDS (Hitachi SU8030; Oxford Aztec X-max 80 SDD EDS detector) were prepared by taking thin-film samples and immersing them in 15 ml DCM in a 3-cm-diameter crystalizing dish.

For samples containing TCBQ, the films delaminated, resulting in small film aggregates that settled to the bottom of the crystalizing dish.

Films were mounted in an upright optical cryostat (Janis Research ST-100) under vacuum ($< 1 \times 10^{-4}$ mbar) in transmission mode in a Cary 5000 UV-Vis-NIR spectrophotometer and spectra were collected from 800–250 nm in 10 K increments.

Emerging Perovskite Monolayers

DOI: https://doi.org/10.1038/s41563-021-01029-9

Originators: Ricciardulli, A.G., Yang, S., Smet, J.H. *et al.* Emerging perovskite monolayers. *Nat. Mater.* 20, 1325–1336 (2021). https://doi.org/10.1038/s41563-021-01029-9

Abstract-Summary

The library of two-dimensional (2D) materials has been enriched over recent years with novel crystal architectures endowed with diverse exciting functionalities.

Bulk perovskites, including metal-halide and oxide systems, provide access to a myriad of properties through molecular engineering.

Emboldened by the 2D materials research, perovskites down to the monolayer limit have recently emerged.

In this Review, we critically assess the preliminary studies on the synthetic routes and inherent properties of monolayer perovskite materials.

Main

Oxide perovskites display diverse physical and chemical properties, which are widely used in ferroelectric, dielectric, piezoelectric, magnetic and energy conversion/ storage devices [54].

Because the organic spacers between two adjacent unit cells of RP-phase halide perovskites are often connected by weak interaction such as van der Waals forces, individual perovskite layers can be isolated from their bulk crystals in the way that graphene has been.

An oxide perovskite monolayer was however successfully exfoliated by ion-exchange methods from DJ-phase bulk materials [55].

We highlight the opportunities to manipulate the properties of perovskite monolayers and offer our insights into the future directions and challenges towards new materials, new device concepts and innovative applications in this emerging field.

Due to the widely tailorable chemical and physical properties, the developments on monolayer perovskites are bound to stimulate fruitful interactions between the currently distinct scientific communities working on perovskites and 2D materials.

Structural versatility

When X is a halide anion (such as Cl^-, Br^- or I^-), A′ is a bulky aromatic or aliphatic alkylammonium spacer, such as butylammonium (BA) and 2-phenylethylammounium (PEA), and B is a divalent metal cation from group XIV (for example, Pb^{2+}, Sn^{2+}, Ge^{2+} and so on) in RP-phase hybrid halide perovskites [56].

The general chemical formula of perovskites is ABX_3, where A and B are both cations and A is generally larger than B, and X is usually an oxide or a halide anion.

Mitzi and co-workers carried out systematic synthesis studies of layered halide perovskite semiconductors with designable organic moieties and tunable electrical properties during the 1990s [32].

It sparked extensive research of halide perovskites as light-absorbing layers in photovoltaic devices.

Layered halide perovskites also offer other interesting photonic, electronic and spin related properties.

Synthesis of perovskite monolayers

Both mechanical exfoliation and MBE usually enable 2D material growth with a clean surface and well-defined morphology, whereas the chemical method based on bottom-up growth in solution often causes surface contamination of the isolated sheets.

Compared with solution-prepared perovskite monolayers, the mechanically exfoliated sheets and those obtained by MBE are more suitable for device assembly and the exploration of new properties.

Compared with rigid layered inorganic materials, the combination of an elastic crystal lattice and weak interlayer interactions makes layered halide perovskites (with mainly $n = 1$) easier to delaminate into thin sheets with large sizes, high crystallinity and surface cleanness.

Although organic molecular beam deposition has been developed for the growth of organic semiconducting layers [57], the different growth mechanism and kinetics for organic layers and inorganic units, and the large lattice mismatch during their sequential growth, strongly hamper the synthesis of hybrid perovskite monolayers.

Tunable electronic and optical properties

These electronic and excitonic properties are easily tuned as the spatial confinement is varied via n. Likewise, the exciton-binding energies of all-inorganic halide perovskites increase dramatically as n decreases ($n \leq 3$).

In molecularly thin $(BA)_2PbBr_4$ sheets, the change in the dielectric environment causes an increase of the exciton binding energy to 490 meV, almost one-third larger compared with the binding energy in the corresponding layered bulk perovskite [30, 58].

While Rashba splitting has been a well-known effect at the interfaces of oxide perovskites, the use of monolayer perovskites instead of the surface of a bulk layer offers additional degrees of freedom and considerably expands the toolbox of the spintronics community.

We note for the sake of completeness that large Rashba splittings not only occur in oxide perovskites but also have been observed in layered halide perovskite with energy splittings as large as (40 ± 5 meV) enriching this field further with the possibility of achieving ferroelectric control over spin textures [59].

Potential device applications

Oxide perovskite monolayers are predestined for the use as catalysts due to their tunable structures and defect chemistry, while halide perovskite monolayers are attractive for photovoltaic applications, although their study for this purpose is at an early stage.

A heterostructure device consisting of a monolayer tungsten disulfide (WS_2) and a thickness-tunable hybrid halide perovskite $(MA)_{n+1}Pb_nI_{3n+1}$ has been shown to exhibit remarkable photosensing and charge separation capabilities [60].

The ORR performance of these materials mostly depends on the surface area, oxide perovskite monolayers with completely exposed active sites can further boost the device performance.

Monolayer halide perovskites are attractive active layers in electrically driven quantum well microcavity devices for polariton laser emission, which operate in the regime of strong coupling between excitons and photons [61].

Additional opportunities in electro-optics and nonlinear optics may arise by tuning the exciton–polariton interaction through chemical composition engineering of monolayer halide perovskites.

Outlook

To the tremendous progress made in bulk perovskite crystals, the journey of exploring monolayer perovskites has just started.

Only a few types of monolayers have been prepared, although numerous perovskites are stable in the 2D regime.

It is crucial to address the challenge of material instability under ambient conditions, especially for the halide perovskite monolayers.

Apart from the required tour de force in synthesis, an intense research thrust to disclose the fundamental physics of perovskite monolayers and the structure–property relationship, and to exploit their potential in prospective, innovative device concepts, is needed.

Hybrid perovskite monolayers are direct-bandgap semiconductors with a 'built in' quantum well confinement potential and a remarkably strong light–matter interaction [62].

Inorganic halide perovskite monolayers form a large family of 2D semiconductors possessing different bandgaps and work functions.

Monolayer-based perovskite devices are rather simplistic.

Deterministic Fabrication of Arbitrary Vertical Heterostructures of Two-Dimensional Ruddlesden–Popper Halide Perovskites

DOI: https://doi.org/10.1038/s41565-020-00802-2

Originators: Pan, D., Fu, Y., Spitha, N. *et al.* Deterministic fabrication of arbitrary vertical heterostructures of two-dimensional Ruddlesden–Popper halide perovskites. *Nat. Nanotechnol.* 16, 159–165 (2021). https://doi.org/10.1038/s41565-020-00802-2

Abstract-Summary

Ruddlesden–Popper lead halide perovskites have emerged as a new class of two-dimensional semiconductors with tunable optoelectronic properties, potentially offering unlimited heterostructure configurations for exploration.

The practical realization of such heterostructures is challenging because of the difficulty in achieving controllable direct synthesis or van der Waals integration of halide perovskites due to their mobile and fragile crystal lattices.

These advances enable the deterministic fabrication of arbitrary vertical heterostructures and multi-heterostructures of Ruddlesden–Popper perovskites with greater structural degrees of freedom that define the electronic structures of the heterojunctions.

Main

Fulfilling the promises of 2D RP perovskite heterostructures, however, demands deterministic fabrication capability to achieve rationally designed electronic structures.

The fragile nature of halide perovskite materials and the limited lateral dimension make transferring and mechanical stacking of exfoliated 2D RP perovskite thin sheets [63] or those directly grown on substrates [39] much less effective and reliable.

Attempts to use direct solution growth of heterostructures of 2D RP perovskites lack control over composition and morphology [64, 65], and have so far only led to mixtures consisting of neighbouring n phases [41, 42, 44, 66].

We report a hybrid approach to deterministically fabricate arbitrary vertical heterostructures of 2D RP lead halide perovskites with different halides, n phases and LA and/or A cations, which is enabled by controllable growth of large-area molecularly thin nanosheets of various phase-pure RP perovskites at the solution–air interface and a facile method for gently transferring and stacking the as-grown sheets.

Floating growth of nanosheets

A warm saturated aqueous solution containing perovskite precursors dissolved in concentrated hydroiodic (or hydrobromic) acid is dropped onto a glass slide.

The droplet becomes supersaturated upon cooling, initiating nucleation and crystal growth.

The key to achieving monolayer or few-layer thicknesses while maximizing lateral growth is maintaining a proper supersaturation, which is carefully optimized by tuning the concentrations of each precursor and the temperatures from which the solution is cooled.

At concentrations greatly exceeding saturation, new nuclei are continuously generated on the existing sheets, which quickly increases the thickness.

At concentrations just above saturation, only a few nuclei are generated, and screw dislocation-driven growth is promoted to produce thick spiral microplates.

Pick-up and transfer of the floating nanosheets

We can pick up these floating thin sheets and transfer them onto arbitrary substrates including glass, quartz and Si/SiO_2 via a soft transfer method without compromising the sheets' structural integrity.

The floating sheets were brought onto the PDMS film while the solution was expelled due to the hydrophobicity of both PDMS and the alkyl tails of the LA ligands.

The perovskite sheets were released and laminated on Si/SiO_2 substrates via elastomeric transfer printing [67].

These thin perovskite sheets are volatile and unstable under intense laser illumination, but their photostability can be substantially improved by coating with a protective polystyrene film.

Deterministic fabrication of arbitrary vertical heterostructures of RP perovskites

The soft crystal lattice of RP perovskites, flexible LA ligands at the surface and the large-area smooth sheets from the floating growth ensure intimate interaction in the stacked van der Waals heterostructures.

The excitonic emission that results in narrow PL emission in RP perovskites is influenced by structural disorder, exciton–phonon interactions and the dielectric environment, all of which are affected by the identity of LA cations and their packing interactions between the soft RP perovskite layers [68].

Previous reports correlated the PL linewidth in $n = 1$ 2D perovskites with molecular rigidity of the LA cations [69, 70].

We postulate that the combination of the more ordered crystal structure due to strong $\pi-\pi$ interaction between adjacent PEA cations and a suitable interlayer distance enabled by the more flexible HA cations may reduce the overall structural and dielectric disorders in the heterostructure, leading to narrower PL emission.

Assembly of complex multi-heterostructures

The capability to assemble arbitrary RP perovskite phases with pristine interfaces enables us to fabricate complex multi-heterostructures with rational control over their electronic structures.

Carrier funnelling from lower n RP phases to higher n phases has been proposed to increase the efficiency in the LEDs made with polycrystalline thin films of those heterostructures [41, 42, 44].

Those junctions are poorly defined because the exact RP phase compositions, stacking sequences and layer orientations are unknown.

Although the detailed carrier transfer mechanism(s) require further investigation and could be dependent on the nature of LA spacer cations, the combination of steady-state PL, TRPL and PLE results strongly suggests that the pristine heterointerfaces present in these vertical heterostructures can efficiently allow carrier transfer between layers.

These clearly defined multi-heterostructures with full control over the compositions (n phase, LA and A cations) and sequences can serve as ideal platforms in which to study the interlayer carrier dynamics.

Conclusions

We report a floating solution growth method for synthesizing large-area atomically thin sheets of diverse 2D RP lead halide perovskites, and show that these floating sheets can be reliably transferred and stacked into 2D vertical heterostructures with clean interfaces.

These advances allow the fabrication of arbitrary vertical heterostructures of RP perovskites with greater degrees of compositional freedom (the halide anions, n value, LA and A cations) that in turn control the electronic structure.

The tunability of the electronic structure (the n value and halide anions) and dielectric environments (LA cations) enables the observation of interlayer carrier transfer and narrowing of PL linewidth, and make these RP perovskite heterostructures a new and versatile platform to explore exciton physics, such as exciton–exciton interactions and Bose–Einstein condensation of excitons [71, 72].

Methods

Confocal PL measurements on the reference thin film and perovskite nanosheets were conducted with a Horiba Labram Aramis confocal PL/Raman microscope using a 442 nm laser.

To selectively excite the n = 3, but not the n = 2, RP perovskite phase, a 590 nm bandpass filter and a 580 nm bandpass filter were combined to yield light at 584 nm (FWHM = 4 nm).

Light was directed to the sample via a ×100 Olympus objective, and PL was collected through the same objective and focused onto the slit of a spectrograph (Princeton Instruments, Acton SP2300i), where it was detected by a 2D array detector (Andor, Neo sCMOS).

A 620 nm bandpass filter was placed in front of the spectrograph entrance slit to isolate the emission from the n = 3 RP perovskite phase.

Layer-by-layer Anionic Diffusion in Two-Dimensional Halide Perovskite Vertical Heterostructures

DOI: https://doi.org/10.1038/s41565-021-00848-w

Originators: Akriti, Shi, E., Shiring, S.B. *et al.* Layer-by-layer anionic diffusion in two-dimensional halide perovskite vertical heterostructures. *Nat. Nanotechnol.* 16, 584–591 (2021). https://doi.org/10.1038/s41565-021-00848-w

Abstract-Summary

Anionic diffusion in a soft crystal lattice of hybrid halide perovskites affects their stability, optoelectronic properties and the resulting device performance.

The use of two-dimensional (2D) halide perovskites improves the chemical stability of perovskites and suppresses the intrinsic anionic diffusion in solid-state devices.

A 'quantized' layer-by-layer diffusion model is proposed to describe the behaviour of the anionic migration in 2D halide perovskites.

Our results provide important insights into the mechanism of anionic diffusion in 2D perovskites and provide a new materials platform with an enhanced stability for heterostructure integration.

Main

Direct visualization of ion migration in three-dimensional (3D) perovskite heterostructures was realized and the solid-state diffusivity of halides is estimated to be in the order of 10^{-14} to 10^{-18} m^2 s^{-1} at a temperature slightly above room temperature, which is dramatically higher than that of Si or oxide perovskites [73–76].

Few years, high-performance optoelectronics with an improved stability and suppressed anion migration were demonstrated using polycrystalline quasi-2D perovskite thin films [14, 77, 78].

A comprehensive quantitative analysis of the factors that impact anionic migration in these 2D halide perovskites is still lacking.

A 'quantized' layer-by-layer diffusion model is proposed to describe the behaviour of the anionic migration in 2D perovskites.

Assembly of vertical heterostructures

Independently synthesizing the perovskite layers before assembling the heterostructure ensures a higher phase purity compared with those of previously reported techniques of heterostructure construction using sequential growth [79] or chemical vapour deposition [65, 80, 81].

Following this synthesis strategy, we constructed a variety of high-quality bromide–iodide 2D perovskite vertical heterostructures.

We first assembled a $(BA)_2PbBr_4$–$(BA)_2(MA)_2Pb_3I_{10}$ vertical heterostructure (BA, butylammonium cation; MA, methylammonium cation).

Confocal PL mapping indicates that the blue emission (the 410–460 nm channel) is limited to the pure bromide region and is largely blocked by the top iodide layer in the junction area due to absorption and/or energy transfer [41, 42, 66, 82–84].

Mechanism of halide diffusion

The emergence of an intermediate green PL emission is an indication of halide diffusion between the top and bottom 2D perovskites of the vertical heterostructure.

This results in a greater impact of interdiffusion on the average composition of bromide perovskite, and thus potentially leads to the formation of an iodide-dominant $(BA)_2PbBr_{4x}I_{4(1-x)}$ alloy with a green PL emission.

An orange-red PL emission was observed after heating the $(BA)_2(MA)_2Pb_3Br_{10}$–$(BA)_2PbI_4$ vertical heterostructure at 80 °C, which is characteristic of an iodide-dominant $(BA)_2(MA)Pb_3Br_{10x}I_{10(1-x)}$ perovskite rather than a bromide-dominant $(BA)_2PbBr_{4x}I_{4(1-x)}$ perovskite.

From the PL spectra evolution during heating of the $(BA)_2PbBr_4$–$(BA)_2(MA)_2Pb_3I_{10}$ vertical heterostructure, it is evident that the iodide-dominant alloy ($(BA)_2PbBr_{4x}I_{4(1-x)}$ with $x \leq 0.5$) forms in the bottom layer within minutes of heating.

The above observations regarding the origin of the green PL emission and formation of the thermodynamically preferred alloy composition provide important insights into the mechanism of halide interdiffusion across the vertical heterojunction.

Diffusion coefficient calculations

The concentration profile at the final time step is fitted with a Gaussian distribution and the diffusion coefficient is then calculated by the Boltzmann–Matano (BM) method [85, 86]: where c^* is the bromide mole fraction in the preferred halide alloy,

c_l is the bromide mole fraction in the top perovskite layer and x_M is the Matano interface (a mathematical parameter that divides the concentration profile into two equal-area plots).

Although the non-Fickian diffusion model is based on the final concentration profile across the vertical heterostructure, it does not capture the unique layer-by-layer alloy-mediated diffusion process observed in the 2D vertical heterostructures.

To model the impact of the layer-by-layer preferred alloy formation on the interdiffusion coefficient, KMC lattice-diffusion simulations were performed.

To capture the experimental behaviour of the heterostructure to equilibrate to a preferred halide alloy composition, as well as to model the proposed layer-by-layer diffusion mechanism, a concentration threshold criterion for the interlayer diffusion was introduced.

Impact of inorganic and organic layers on halide diffusion

All the models predicted that the halide interdiffusion increases with the inorganic layer thickness (n), which is consistent with the reported bromide–iodide interdiffusion coefficient of ~10^{-18} m^2 s^{-1} for 3D perovskite heterostructure [74, 76].

For a given organic cation, 2D perovskites ($n = 1$) are better inhibitors of halide diffusion compared with quasi-2D ($n > 1$) and 3D perovskites.

The effect of an organic cation on the halide interdiffusion in 2D perovskites was investigated by comparing BA with three larger cations—phenylethylammonium (PEA), biphenylethylammonium (2P) and bithiophenylethylammonium (2T) [87].

All the diffusion models present a consistent trend, which indicates that bulkier π-conjugated organic cations, such as 2T, are much more efficient in inhibiting halide interdiffusion compared with benzene-based π-conjugated systems, such as 2P and PEA, or short-chain aliphatic organic cations, such as BA.

Conclusions

We have demonstrated the assembly of atomically flat vertical heterostructures using two 2D halide perovskites.

To classical solid-state interdiffusion models, which feature a continuous concentration profile evolution, a quantized layer-by-layer diffusion model governed by a concentration threshold and ion-blocking effects of the organic cations was discovered in 2D halide perovskites.

The realization of vertical heterostructures also provides a materials platform for fundamental studies on the interplay between different 2D halide perovskites and their optoelectronic applications.

Methods

For the synthesis of $(BA)_2PbBr_4$ nanocrystals, a concentrated precursor solution was prepared by dissolving 0.02 mmol n-butylammonium bromide (BA·HBr) and 0.01 mmol lead bromide ($PbBr_2$) in a 2 ml solution of dimethylformamide (DMF) and chlorobenzene (CB) (DMF:CB volume ratio of 1:1).

For the synthesis of $(BA)_2PbI_4$ bulk crystals, a precursor solution was prepared by mixing 0.57 mmol n-butylammonium iodide (BA·HI) and 0.57 mmol PbO in 0.9 ml of hydroiodic acid (HI, 57 wt% in H_2O) and 0.1 ml of H_3PO_2 (50 wt% in H_2O).

The perovskite crystals were obtained by a controlled slow cooling at a rate of ~5 °C h^{-1} followed by vacuum filtration. $(BA)_2(MA)_2Pb_3I_{10}$ bulk crystals were synthesized using a solution of 0.19 mmol BA·HI, 0.40 mmol MA·HI and 0.59 mmol PbO in 0.9 ml of HI (57 wt% in H_2O) and 0.1 ml of H_3PO_2 (50 wt% in H_2O).

A Review of Main Characterization Methods for Identifying Two-Dimensional Organic–Inorganic Halide Perovskites

DOI: https://doi.org/10.1007/s10853-021-06038-2

Originators: Yu, S., Liu, P. & Xiao, S. A review of main characterization methods for identifying two-dimensional organic–inorganic halide perovskites. *J Mater Sci* 56, 11656–11681 (2021). https://doi.org/10.1007/s10853-021-06038-2

Abstract-Summary

Two-dimensional (2D) halide perovskite has attracted widespread attention because of their structural diversity and adjustable material properties.

Their material properties are closely related to structural features, especially the inorganic layers (n) in each structural unit of 2D halide perovskite.

This review introduces the basic structure and properties of 2D halide perovskite and then summarizes various experimental methods that can be used to directly or indirectly identify the n value, such as X-ray diffraction (XRD), grazing incidence wide-angle X-ray scattering (GIWAXS), photoluminescence spectra (PL), ultraviolet–visible spectra (UV–vis), atomic force microscopy (AFM), second harmonic generation (SHG) spectroscopy or third harmonic generation (THG) spectroscopy, etc Moreover, the existence of different n-value phases in 2D halide perovskites has been discussed in terms of the energy transfer characteristics.

Introduction

The investigation of hybrid organic–inorganic halide perovskite in the field of solar cells has achieved remarkable improvement in the power conversion efficiency (PCE) in the past decade.

Although perovskite materials have achieved high efficiency comparable to commercial solar cell devices, their environmental stability is still far from meeting industrial requirements.

Layered perovskite solar cell was first demonstrated by Karunadasa and coworkers [88] in 2014 using $(C_6H_5(CH_2)_2NH_3)_2(CH_3NH_3)_2Pb_3I_{10}$ perovskite and with a PCE of 4.73%.

Hybrid 2D perovskite in DJ structure was crystallized and implemented in solar cells for the first time by Kanatzidis and coworkers using 3- or 4-(ammoniomethyl) piperidinium and MA in 2018 [89].

The general concept and material properties are provided for 2D hybrid perovskite.

2D structures and properties

2D halide perovskite can be generally understood as sandwich structure, in which BX_6 inorganic octahedron layer is sandwiched between two organic spacer cations.

Organic spacer cations are composed of monovalent or divalent organic amines which are different from cations at A-site, and write as C. If the material is composed of a single layer of BX_6 inorganic octahedrons closely arranged, it is called pure 2D perovskite.

The distance between 2D perovskite inorganic layers can be controlled by selecting different organic spacer cations C [32, 90].

The tolerance for such a wide variety of sizes and functional groups in the organic cations, along with the possibility to change the metallic, halide atoms, and the thickness of the inorganic sheets (n value), makes 2D hybrid perovskite virtually infinite, and yet, only a few hundred perovskite structures have been reported.

2D or not? Characterization methods

The crystal structure and optical properties of 2D perovskite which vary with the n value can be verified by the following characterization methods: the growth of crystal and the stacking way are verified by XRD and GIWAXS; PL and UV–vis characterize the energy level arrangement structure and explore the optical properties of perovskite crystals; AFM is mainly used to explore the surface morphology of the crystal; SHG/THG is a unique characterization method for 2D perovskite to study nonlinear optical properties.

For other types of 2D perovskite in RP structure, the trend of the UV–vis absorption spectra and PL spectra with respect to the n value is the same as the trend of PEA and BA cations, but the degrees of the changes of the intensity value and the positions of the absorption and emission peaks are different.

Energy band determined by structure

A typical 2D layered perovskite is composed of a mixture of 2D perovskite materials of different n values, in which energy is rapidly transferred from a low n value layer (donor) to a high n value layer (acceptor).

The energy transfer between different n-value phases reduces the non-radiative recombination of excitons and improves the photovoltaic efficiencies of 2D perovskite-based solar cells.

The 2D layered perovskite can be systematically regulated for the Stokes displacement, that is, by adjusting the ratio of different n-value perovskite for the light absorption layer to reduce the energy loss, in order to maintain a higher photoluminescence quantum yield (PLQY).

These QWs with different n values establish an energy cascade for exciton transfer (ET) from a small n-value wide band gap QW to a large n-value narrow band gap QW [91].

Conclusion and outlook

Different n values of 2D perovskites determine that they have different energy band structures.

The addition of C and inorganic layer make the 2D perovskite form a quantum well structure, and physical properties therefore have become diverse.

There are many steps at the edges of 2D perovskite crystals, which can be characterized by AFM to obtain n value.

The alternate arrangement of 2D perovskite organic layer and inorganic layer will form a regular crystal plane reflection peak that is different from 3D perovskite.

An effective balance of stability (under light and heat) and photoelectric efficiency can be achieved by exploring the energy transfer between 3D/2D perovskites. (2) Unveiling the inner mechanism of optical properties of 2D perovskite.

It has broad application prospects in the field of optoelectronics, especially high-energy X-ray detection and laser technology. (4) Accurately characterizing the distribution of n value in 2D perovskites.

Ruddlesden–Popper 2D Perovskites of Type $(C_6H_9C_2H_4NH_3)_2(CH_3NH_3)_{n-1}Pb_nI_{3n+1}$ (n = 1–4) for Optoelectronic Applications

DOI: https://doi.org/10.1038/s41598-022-06108-8

This article is licensed under a Creative Commons Attribution 4.0 International License, which permits use, sharing, adaptation, distribution and reproduction in any medium or format, as long as you give appropriate credit to the original author(s) and the source, provide a link to the Creative Commons license, and indicate if changes were made. The images or other third party material in this article are included in the article's Creative Commons license, unless indicated otherwise in a credit line to the material. If material is not included in the article's Creative Commons license and your intended use is not permitted by statutory regulation or exceeds the permitted use, you will need to obtain permission directly from the copyright holder. To view a copy of this license, visit http://creativecommons.org/licenses/by/4.0/.

Originators: Rahil, M., Ansari, R.M., Prakash, C. *et al.* Ruddlesden–Popper 2D perovskites of type (C6H9C2H4NH3)2(CH3NH3)n−1PbnI3n+1 (n = 1–4) for opto-electronic applications. *Sci Rep* 12, 2176 (2022). https://doi.org/10.1038/s41598-022-06108-8

Abstract-Summary

Ruddlesden–Popper (RP) phase metal halide organo perovskites are being extensively studied due to their quasi-two dimensional (2D) nature which makes them an excellent material for several optoelectronic device applications such as solar cells, photo-detectors, light emitting diodes (LEDs), lasers etc While most of reports show use of linear carbon chain based organic moiety, such as n-Butylamine, as organic spacer in RP perovskite crystal structure, here we report a new series of quasi 2D perovskites with a ring type cyclic carbon group as organic spacer forming RP perovskite of type $(CH)_2(MA)_{n-1}Pb_nI_{3n+1}$; CH = 2-(1-Cyclohexenyl)ethylamine; MA = Methylamine).

The demonstrated RP perovskite of type for n = 1–4 have shown formation of highly crystalline thin films with alternate stacking of organic and inorganic layers, where the order of PbI_6 octahedron layering are controlled by n-value, and shown uniform direct bandgap tunable from 2.51 eV (n = 1) to 1.92 eV (n = 4).

Thermogravimetric analysis (TGA) showed highly stable nature of reported RP perovskites with linear increase in phase transition temperatures from 257 °C (n = 1) to 270 °C (n = 4).

The photodetectors fabricated for the series using $(CH)_2(MA)_{n-1}Pb_nI_{3n+1}$ RP perovskite as active absorbing layer and without any charge transport layers, shown sharp photocurrent response from 17 nA/cm^2 for n = 1 to 70 nA/cm^2 for n = 4, under zero bias and low power illumination conditions (470 nm LED, 1.5 mW/cm^2).

For lowest bandgap RP perovskite n = 4, $(CH)_2MA_3Pb_4I_{13}$ the photodetector showed maximum photocurrent density of ~ 508 nA/cm^2 at 3 V under similar illumination condition, thus giving fairly large responsivity (46.65 mA/W).

Introduction

Apart from electroluminescent devices, in 2016 Zhou and others, reported first RP metal halide perovskite based photodetector, fabricated in lateral two electrode configuration, with $(BA)_2(MA)_{n-1}Pb_nI_{3n+1}$ (n = 1–3) as the absorbing layer [92].

In 2019 Liu and others, represented the effect of tailoring the large organic spacer cation on the optoelectronic properties of RP lead iodide n = 1 perovskites [93] by fabricating $(BA)_2PbI_4$ (BA: butylamine), $(HA)_2PbI_4$ (HA: hexylamine) and $(OA)_2PbI_4$ (OA: octylamine) and investigated their optical, structural, and optoelectronic properties.

We are reporting the synthesis and optoelectronic studies of new quasi-2D perovskites based on 2-(1-Cyclohexenyl)ethylamine ($C_6H_9C_2H_4NH_2^+$, hereafter CH) as organic spacer in the RP phase lead iodide perovskite of type $(CH)_2(MA)_{n-1}Pb_nI_{3n+1}$ with n = 1–4, namely $(CH)_2PbI_4$ (n = 1), $(CH)_2(MA)Pb_2I_7$ (n = 2), $(CH)_2(MA)_2Pb_3I_{10}$ (n = 3), and $(CH)_2(MA)_3Pb_4I_{13}$ (n = 4).

Results and discussion

For n = 2 RP perovskite, with the introduction of methylamine (MA), stoichiometric formula converts to $(CH)_2MAPb_2I_7$ and intense diffraction peak corresponding to (0k0) planes were observed, showing the presence of 3D perovskite phase [94–96].

For n = 2 $(CH)_2(MA)Pb_2I_7$ film, an exciton absorption peak is observed at 567 nm (λ_2) along with the characteristic absorbance peak corresponding to n = 1 at 508 nm (λ_1) thus showing the presence of lower n value phase in the RP perovskites [97].

Small values of Stokes' shift ranging between 50 and 30 meV for n = 1–4 thin films shows the formation of highly defect free RP phase perovskite structures, which results in reduced non-radiative recombination losses, making these solution processed RP perovskite semiconductors a possible candidate for optoelectronic devices such as light-emitting diodes, lasers etc The charge carrier recombination lifetimes are measured for n = 1–4 thin films, spin coated on glass substrates, using the time correlated single photon counting (TCSPC) method.

Conclusion

The cyclic organic spacer in the RP perovskite structure has improved the confinement effects which has enabled the formation of sharp and intense PL at room temperature for all thin films (n = 1–4).

Transient photocurrent measurements performed on photodetectors with $(CH)_2(MA)_{n-1}Pb_nI_{3n+1}$ (n = 1–4) RP perovskite as light absorbing layer showed dependency of photocurrent on the number of inorganic sheets and applied bias.

n = 4 perovskite showed highest photocurrent compare than lower member of series reaching 508 nA/cm^2 at 3 V. Overall the cyclic ring type organic spacer cation CH based RP perovskites has shown improved optoelectronic properties such as tunable energy band gaps, high carrier mobilities and charge carrier lifetimes, as well as promising structural stability which makes them a potential candidate to be explored for the development of advanced optoelectronic devices such as photodetector, solar cells, LEDs, photobatteries etc

Experimental Section

After the hot PbI_2 solution turned into bright yellow solution, MACl (2.5 mmol for $(CH)_2(MA)PbI_7$ (n = 2), 3.33 mmol for $(CH)_2(MA)_2Pb_3I_{10}$ (n = 3) and 3.75 mmol for $(CH)_2(MA)_3Pb_4I_{13}$ (n = 4)) is added subsequently and the yellow solution showed the black precipitate formation, which were redissolved under stirring and reflux condition at heating to boiling.

The $2\text{-}C_6H_9C_2H_4NH_3I$ solution is mixed with the hot PbI_2 solution under constant stirring and heating and 7.5 mmol 57% w/w aqueous HI solution is added for proper mixing.

The glass substrate is heated to 130 °C and the DMF solution of perovskite is heated to 70 °C, and spin-coating is performed at 1000 rpm for 5 s by quickly mounting the glass substrate on the spin coater chuck, followed by a drying step of 3000 rpm for 40 s. The resultants thin films were annealed at 100 °C for 10 min on the hot plate.

Bibliography

1. Billing, D.G., Lemmerer, A.: Synthesis, characterization and phase transitions in the inorganic–organic layered perovskite-type hybrids [$(C_nH_{2n+1}NH_3)_2PbI_4$], n = 4, 5 and 6. Acta Crystallogr. **B 63**, 735–747 (2007) (Description of the detailed structural changes taking place during phase transitions in 2DPKs with n = 1)
2. Barman, S., Venkataraman, N.V., Vasudevan, S., Seshadri, R.: Phase transitions in the anchored organic bilayers of long-chain alkylammonium lead iodides ($(C_nH_{2n+1}NH_3)_2PbI_4$; n = 12, 16, 18. J. Phys. Chem. B **107**, 1875–1883 (2003)
3. Spanopoulos, I., et al.: Uniaxial expansion of the 2D Ruddlesden-Popper perovskite family for improved environmental stability. J. Am. Chem. Soc. **141**, 5518–5534 (2019)
4. Soe, C.M.M., et al.: Structural and thermodynamic limits of layer thickness in 2D halide perovskites. Proc. Natl. Acad. Sci. **116**, 58–66 (2019)
5. Liu, Y., Xiao, H., Goddard, W.A.: Two-dimensional halide perovskites: tuning electronic activities of defects. Nano Lett. **16**, 3335–3340 (2016)
6. Pandey, M., Jacobsen, K.W., Thygesen, K.S.: Band gap tuning and defect tolerance of atomically thin two-dimensional organic–inorganic halide perovskites. J. Phys. Chem. Lett. **7**, 4346–4352 (2016)
7. Kitazawa, N., Aono, M., Watanabe, Y.: Synthesis and luminescence properties of lead-halide based organic-inorganic layered perovskite compounds $(C_nH_{2n+1}NH_3)_2PbI_4$ (n=4, 5, 7, 8 and 9). J. Phys. Chem. Solids **72**, 1467–1471 (2011)
8. Blancon, J.-C., et al.: Extremely efficient internal exciton dissociation through edge states in layered 2D perovskites. Science **355**, 1288–1292 (2017)
9. Wang, Y.-K., et al.: Chelating-agent-assisted control of $CsPbBr_3$ quantum well growth enables stable blue perovskite emitters. Nat. Commun. **11**, 3674 (2020)
10. Tsai, H., et al.: Stable light-emitting diodes using phase-pure Ruddlesden-Popper layered perovskites. Adv. Mater. **30**, 1704217 (2018)
11. Yang, A., et al.: Giant enhancement of photoluminescence emission in WS2-two-dimensional perovskite heterostructures. Nano Lett. **19**, 4852–4860 (2019)
12. Quan, L.N., et al.: Edge stabilization in reduced-dimensional perovskites. Nat. Commun. **11**, 1–9 (2020)
13. Shang, Y. et al. Highly stable hybrid perovskite light-emitting diodes based on Dion-Jacobson structure. Sci. Adv. 5, eaaw8072 (2019).
14. Cao, D.H., Stoumpos, C.C., Farha, O.K., Hupp, J.T., Kanatzidis, M.G.: 2D homologous perovskites as light-absorbing materials for solar cell applications. J. Am. Chem. Soc. **137**, 7843–7850 (2015)
15. Stoumpos, C.C., et al.: Ruddlesden-Popper hybrid lead iodide perovskite 2D homologous semiconductors. Chem. Mater. **28**, 2852–2867 (2016)
16. Du, K., et al.: Two-dimensional lead(ii) halide-based hybrid perovskites templated by acene alkylamines: crystal structures, optical properties, and piezoelectricity. Inorg. Chem. **56**, 9291–9302 (2017)
17. Tsai, H., et al.: High-efficiency two-dimensional Ruddlesden-Popper perovskite solar cells. Nature **536**, 312–316 (2016)
18. Wu, X., Trinh, M.T., Zhu, X.-Y.: Excitonic many-body interactions in two-dimensional lead iodide perovskite quantum wells. J. Phys. Chem. **119**, 14714–14721 (2015)
19. Grancini, G., et al.: Role of microstructure in the electron–hole interaction of hybrid lead halide perovskites. Nat. Photon. **9**, 695–701 (2015)
20. Cortecchia, D., et al.: Polaron self-localization in white-light emitting hybrid perovskites. J. Mater. Chem. C **5**, 2771–2780 (2017)
21. Hu, T., et al.: Mechanism for broadband white-light emission from two-dimensional (110) hybrid perovskites. J. Phys. Chem. Lett. **7**, 2258–2263 (2016)
22. Neogi, I., et al.: Broadband-emitting 2D hybrid organic–inorganic perovskite based on cyclohexane-bis(methylamonium) cation. Chemsuschem **10**, 3765–3772 (2017)

23. Yangui, A., et al.: Optical investigation of broadband white-light emission in self-assembled organic–inorganic perovskite $(C_6H_{11}NH_3)_2PbBr_4$. J. Phys. Chem. **119**, 23638–23647 (2015)
24. Grätzel, M.: The rise of highly efficient and stable perovskite solar cells. Acc. Chem. Res. **50**, 487–491 (2017)
25. Leijtens, T., et al.: Overcoming ultraviolet light instability of sensitized TiO_2 with meso-superstructured organometal tri-halide perovskite solar cells. Nat. Commun. **4**, 2885 (2013)
26. Domanski, K., et al.: Not all that glitters is gold: metal-migration-induced degradation in perovskite solar cells. ACS Nano **10**, 6306–6314 (2016)
27. Yao, K., Wang, X., Li, F., Zhou, L.: Mixed perovskite based on methyl-ammonium and polymeric-ammonium for stable and reproducible solar cells. Chem. Commun. **51**, 15430–15433 (2015)
28. Yao, K., Wang, X., Xu, Y., Li, F., Zhou, L.: Multilayered perovskite materials based on polymeric-ammonium cations for stable large-area solar cell. Chem. Mater. **28**, 3131–3138 (2016)
29. Gélvez-Rueda, M.C., et al.: Interconversion between free charges and bound excitons in 2D hybrid lead halide perovskites. J. Phys. Chem. C **121**, 26566–26574 (2017)
30. Ishihara, T., Takahashi, J., Goto, T.: Exciton state in two-dimensional perovskite semiconductor $(C_{10}H_{21}NH_3)_2PbI_4$. Solid State Commun. **69**, 933–936 (1989)
31. Meresse, A., Daoud, A.: Bis(n-propylammonium) tetrachloroplumbate. Acta Crystallogr. C **45**, 194–196 (1989)
32. Mitzi, D.B., Feild, C.A., Harrison, W.T.A., Guloy, A.M.: Conducting tin halides with a layered organic-based perovskite structure. Nature **369**, 467–469 (1994)
33. Mitzi, D.B., Wang, S., Feild, C.A., Chess, C.A., Guloy, A.M.: Conducting layered organic-inorganic halides containing <110>-oriented perovskite sheets. Science **267**, 1473–1476 (1995)
34. Papavassiliou, G.C.: Three- and low-dimensional inorganic semiconductors. Prog. Solid State Chem. **25**, 125–270 (1997)
35. Kagan, C.R., Mitzi, D.B., Dimitrakopoulos, C.D.: Organic–inorganic hybrid materials as semiconducting channels in thin-film field-effect transistors. Science **286**, 945–947 (1999)
36. Borriello, I., Cantele, G., Ninno, D.: Ab initio investigation of hybrid organic-inorganic perovskites based on tin halides. Phys. Rev. B **77**, 235214 (2008)
37. Lanty, G., et al.: Room-temperature optical tunability and inhomogeneous broadening in 2D-layered organic–inorganic perovskite pseudobinary alloys. J. Phys. Chem. Lett. **5**, 3958–3963 (2014)
38. Smith, I.C., Hoke, E.T., Solis-Ibarra, D., McGehee, M.D., Karunadasa, H.I.: A layered hybrid perovskite solar-cell absorber with enhanced moisture stability. Angew. Chem. Int. Ed. **53**, 11232–11235 (2014)
39. Dou, L.T., et al.: Atomically thin two-dimensional organic-inorganic hybrid perovskites. Science **349**, 1518–1521 (2015)
40. Ha, S.T., Shen, C., Zhang, J., Xiong, Q.H.: Laser cooling of organic-inorganic lead halide perovskites. Nat. Photon. **10**, 115–121 (2016)
41. Wang, N., et al.: Perovskite light-emitting diodes based on solution-processed self-organized multiple quantum wells. Nat. Photon. **10**, 699–704 (2016)
42. Yuan, M., et al.: Perovskite energy funnels for efficient light-emitting diodes. Nat. Nanotechnol. **11**, 872–877 (2016)
43. Mao, L.L., et al.: Tunable white-light emission in single-cation-templated three-layered 2D perovskites $(CH_3CH_2NH_3)_4Pb_3Br_{10-x}Cl_x$. J. Am. Chem. Soc. **139**, 11956–11963 (2017)
44. Proppe, A.H., et al.: Synthetic control over quantum well width distribution and carrier migration in low-dimensional perovskite photovoltaics. J. Am. Chem. Soc. **140**, 2890–2896 (2018)
45. Shi, E., et al.: Two-dimensional halide perovskite nanomaterials and heterostructures. Chem. Soc. Rev. **47**, 6046–6072 (2018)
46. Mitzi, D.B.: Templating and structural engineering in organic–inorganic perovskites. J. Chem. Soc. Dalton Trans. 1–12 (2001)

47. Saparov, B., Mitzi, D.B.: Organic–inorganic perovskites: structural versatility for functional materials design. Chem. Rev. **116**, 4558–4596 (2016)
48. Evans, H.A., et al.: (TTF)Pb2I5: a radical cation-stabilized hybrid lead iodide with synergistic optoelectronic signatures. Chem. Mater. **28**, 3607–3611 (2016)
49. Mitzi, D.B., Chondroudis, K., Kagan, C.R.: Design, structure, and optical properties of organic–inorganic perovskites containing an oligothiophene chromophore. Inorg. Chem. **38**, 6246–6256 (1999)
50. Mitzi, D.B., Medeiros, D.R., Malenfant, P.R.L.: Intercalated organic−inorganic perovskites stabilized by fluoroaryl−aryl interactions. Inorg. Chem. **41**, 2134–2145 (2002)
51. Lemmerer, A., Billing, D.G.: Synthesis, characterization and phase transitions of the inorganic–organic layered perovskite-type hybrids [$(CnH_{2n}+1NH_3)_2PbI_4$], n = 7, 8, 9 and 10. Dalton Trans. **41**, 1146–1157 (2012)
52. Billing, D.G., Lemmerer, A.: Synthesis, characterization and phase transitions in the inorganic–organic layered perovskite-type hybrids $(CnH_{2n}+1NH_3)_2PbI_4$, n = 4, 5 and 6. Acta Crystallogr. B **63**, 735–747 (2007)
53. Passarelli, J.V., et al.: Enhanced out-of-plane conductivity and photovoltaic performance in n = 1 layered perovskites through organic cation design. J. Am. Chem. Soc. **140**, 7313–7323 (2018)
54. Sun, C., Alonso, J.A., Bian, J.: Recent advances in perovskite-type oxides for energy conversion and storage applications. Adv. Energy Mater. **11**, 2000459 (2021)
55. Li, S., Zhang, Y., Yang, W., Liu, H., Fang, X.: 2D perovskite $Sr_2Nb_3O_{10}$ for high-performance UV photodetectors. Adv. Mater. **32**, 1905443 (2020)
56. Chen, Y., et al.: 2D Ruddlesden-Popper perovskites for optoelectronics. Adv. Mater. **30**, 1703487 (2018)
57. Kowarik, S., Gerlach, A., Schreiber, F.: Organic molecular beam deposition: fundamentals, growth dynamics, andin situstudies. J. Phys. Condens. Matter **20**, 184005 (2008)
58. Yaffe, O., et al.: Excitons in ultrathin organic–inorganic perovskite crystals. Phys. Rev. B **92**, 045414 (2015)
59. Zhai, Y., et al.: Giant Rashba splitting in 2D organic–inorganic halide perovskites measured by transient spectroscopies. Sci. Adv. **3**, e1700704 (2017)
60. Yang, T., et al.: Ultrahigh-performance optoelectronics demonstrated in ultrathin perovskite-based vertical semiconductor heterostructures. ACS Nano **13**, 7996–8003 (2019)
61. Schneider, C., et al.: An electrically pumped polariton laser. Nature **497**, 348–352 (2013)
62. Qi, X., et al.: Photonics and optoelectronics of 2D metal-halide perovskites. Small **14**, 1800682 (2018)
63. Leng, K., et al.: Molecularly thin two-dimensional hybrid perovskites with tunable optoelectronic properties due to reversible surface relaxation. Nat. Mater. **17**, 908–914 (2018)
64. Wang, J., et al.: Controllable synthesis of two-dimensional Ruddlesden-Popper-type perovskite heterostructures. J. Phys. Chem. Lett. **8**, 6211–6219 (2017)
65. Wang, J., et al.: Controllable growth of centimeter-sized 2D perovskite heterostructures for highly narrow dual-band photodetectors. ACS Nano **13**, 5473–5484 (2019)
66. Fu, Y., et al.: Multicolor heterostructures of two-dimensional layered halide perovskites that show interlayer energy transfer. J. Am. Chem. Soc. **140**, 15675–15683 (2018)
67. Meitl, M.A., et al.: Transfer printing by kinetic control of adhesion to an elastomeric stamp. Nat. Mater. **5**, 33–38 (2005)
68. Mauck, C.M., Tisdale, W.A.: Excitons in 2D organic–inorganic halide perovskites. Trends Chem. **1**, 380–393 (2019)
69. Gong, X.W., et al.: Electron-phonon interaction in efficient perovskite blue emitters. Nat. Mater. **17**, 550–556 (2018)
70. Zhang, S., et al.: Synthesis and optical properties of novel organic–inorganic hybrid nanolayer structure semiconductors. Acta Mater. **57**, 3301–3309 (2009)
71. Butov, L.V., Zrenner, A., Abstreiter, G., Bohm, G., Weimann, G.: Condensation of indirect excitons in coupled AlAs/GaAs quantum wells. Phys. Rev. Lett. **73**, 304–307 (1994)

72. Butov, L.V., et al.: Stimulated scattering of indirect excitons in coupled quantum wells: signature of a degenerate Bose-gas of excitons. Phys. Rev. Lett. **86**, 5608–5611 (2001)
73. Pan, D., et al.: Visualization and studies of ion-diffusion kinetics in cesium lead bromide perovskite nanowires. Nano Lett. **18**, 1807–1813 (2018)
74. Zhang, Y., et al.: Quantitative imaging of anion exchange kinetics in halide perovskites. Proc. Natl. Acad. Sci. USA **116**, 12648–12653 (2019)
75. Lai, M., et al.: Intrinsic anion diffusivity in lead halide perovskites is facilitated by a soft lattice. Proc. Natl. Acad. Sci. USA **115**, 11929–11934 (2018)
76. Shewmon, N.T., Yu, H., Constantinou, I., Klump, E., So, F.: Formation of perovskite heterostructures by ion exchange. ACS Appl. Mater. Inter. **8**, 33273–33279 (2016)
77. Smith, I.C., Hoke, E.T., Solis-Ibarra, D., McGehee, M.D., Karunadasa, H.I.: A layered hybrid perovskite solar cell absorber with enhanced moisture stability. Angew. Chem. Int. Ed. **53**, 11232–11235 (2014)
78. Saidaminov, M.I., Mohammed, O.F., Bakr, O.M.: Low-dimensional-networked metal halide perovskites: the next big thing. ACS Energy Lett. **2**, 889–896 (2017)
79. Shi, E., et al.: Two-dimensional halide perovskite lateral epitaxial heterostructures. Nature **580**, 614–620 (2020)
80. Wang, J., et al.: Controllable synthesis of two-dimensional Ruddlesden–Popper-type perovskite heterostructures. J. Phys. Chem. Lett. **8**, 6211–6219 (2017)
81. Hwang, B.H., Lee, J.-S.: 2D perovskite-based self-aligned lateral heterostructure photodetectors utilizing vapor deposition. Adv. Opt. Mater. **7**, 1801356 (2019)
82. Xing, G., et al.: Transcending the slow bimolecular recombination in lead–halide perovskites for electroluminescence. Nat. Commun. **8**, 14558 (2017)
83. Quan, L.N., et al.: Tailoring the energy landscape in quasi-2D halide perovskites enables efficient green-light emission. Nano Lett. **17**, 3701–3709 (2017)
84. Shang, Q., et al.: Unveiling structurally engineered carrier dynamics in hybrid quasi-two-dimensional perovskite thin films toward controllable emission. J. Phys. Chem. Lett. **8**, 4431–4438 (2017)
85. Tuck, B.: Atomic Diffusion in III–V Semiconductors Ch 2 (IOP, 1988)
86. Matano, C.: On the relation between the diffusion-coefficients and concentrations of solid metals (the nickel–copper system). Jpn J. Phys. **8**, 109–113 (1993)
87. Gao, Y., et al.: Molecular engineering of organic–inorganic hybrid perovskites quantum wells. Nat. Chem. **11**, 1151–1157 (2019)
88. Smith, I.C., Hoke, E.T., Solis-Ibarra, D., McGehee, M.D., Karunadasa, H.I.: A layered hybrid perovskite solar-cell absorber with enhanced moisture stability. Angew. Chem. Int. Ed. **53**(42), 11232–11235 (2014)
89. Mao, L., et al.: Hybrid Dion-Jacobson 2D lead iodide perovskites. J. Am. Chem. Soc. **140**, 3775–3783 (2018)
90. Calabrese, J., Jones, N.L., Harlow, R.L., Herron, N., Thorn, D.L., Wang, Y.: Preparation and characterization of layered lead halide compounds. J. Am. Chem. Soc. **113**(6), 2328–2330 (1991). https://doi.org/10.1021/ja00006a076
91. Li, M., Gao, Q., Liu, P., Liao, Q., Zhang, H., Yao, J., Hu, W., Wu, Y., Fu, H.: Amplified spontaneous emission based on 2D ruddlesden-popper perovskites. Adv. Funct. Mater. **28**(17), 1707006 (2018). https://doi.org/10.1002/adfm.201707006
92. Zhou, J., Chu, Y., Huang, J.: Photodetectors based on two-dimensional layer-structured hybrid lead iodide Perovskite semiconductors. ACS Appl. Mater. Inter. **8**, 25660–25666 (2016)
93. Liu, G., et al.: Influence of the organic chain on the optical properties of two-dimensional organic-inorganic hybrid lead iodide perovskites. ACS Appl. Electron. Mater. **1**, 2253–2259 (2019)
94. Myae Soe, C.M., et al.: Structural and thermodynamic limits of layer thickness in 2D halide perovskites. Proc. Natl. Acad. Sci. U.S.A. 116, 58–66 (2019)
95. Mao, L., Stoumpos, C.C., Kanatzidis, M.G.: Two-dimensional hybrid halide perovskites: principles and promises. J. Am. Chem. Soc. **141**, 1171–1190 (2019)

96. Raghavan, C.M., et al.: Low-threshold lasing from 2D homologous organic-inorganic hybrid Ruddlesden-Popper perovskite single crystals. Nano Lett. **18**, 3221–3228 (2018)
97. Zhang, Y., et al.: Two-dimensional (PEA)2PbBr$_4$ perovskite single crystals for a high performance UV-detector. J. Mater. Chem. C **7**, 1584–1591 (2019)

Chapter 2
Optical and Electronic Properties

Introduction by Editor

In this chapter we present many details on the optical and electronic properties of 2D metal halide perovskites (MHPs). As mentioned previously, the 2D MHP structure can be thought of as a quantum well superlattice which includes effects from both quantum and dielectric confinement. The lower bandgaps in the metal halide layers form the quantum wells which tend to dominate the electronic structure, and the organic cations serve as the barriers. The dielectric confinement comes from a mismatch in dielectric constant between the wells and barriers, with the organic cations having lower dielectric constants than the inorganic layers. In addition, the confinement of electrons and holes to the same 2D layer creates a substantially enhanced excitonic effect compared to 3D perovskites. Thus, excitons play a central role in dictating the 2D MHP optoelectronic properties, which in turn impacts their performance in applications, from optical absorption to charge transport to nonradiative Auger recombination.

We present the papers in this chapter in chronological order. They deal with a wide range of situations related to these electronic properties. The large enhancement of exciton binding energies in 2D MHPs (typically hundreds of meVs, compared to tens of meVs for 3D MHPs) tends to limit their applications in optoelectronic devices that require charge separation, such as solar cells and photodetectors. Thus much research has been done on how to alter the exciton binding energies. One way of dealing with this is to control properties through composition. For example, a high dielectric constant organic layer leads to less of a dielectric mismatch and a corresponding lowered exciton binding energy, making such MHPs potentially more useful for charge separating applications. Special organic layers can also be used which directly contribute to the electronic bands, as opposed to simply serving as barriers. Also affecting both exciton binding energy as well as the overall bandgap

© The Author(s), under exclusive license to Springer Nature Singapore Pte Ltd. 2024
J. S. Colton and K. R. Hansen (eds.), *Two-Dimensional Metal Halide Perovskites*,
https://doi.org/10.1007/978-981-99-7830-4_2

is the so-called "n number", i.e., the number of metal halide monolayers that form between the organic barriers. This number can be adjusted through growth methods, but also can be dynamically altered through exposure to vapor or by changing the temperature. The presence of light can also dynamically affect the electronic energies in atomically thin 2D MHPs.

Regarding electronic transport, the orientation of the planes in the structure, e.g. parallel vs perpendicular to the substrate leads to important differences in transport behavior which can be very important in device performance. Defects of course play their own role in affecting transport, as does excitonic diffusion. With regards to the latter, both exciton-phonon scattering and exciton-defect scattering are important, with large differences in their relative importance occurring in Sn-based compounds as compared to the more traditional Pb-based compounds. Meanwhile, from a theoretical perspective, density functional theory is the method nearly universally used to calculate the electronic band structures, and therefore we have chosen a representative paper to discuss this for a specific 2D MHP compound. Finally, doping being of critical importance for nearly any semiconductor system, we also include a review of doping in MHPs, with both intrinsic and extrinsic dopants.

Machine generated summaries

Disclaimer: The auto-summaries have been generated by a recursive clustering algorithm via the Dimensions Auto-summarizer by Digital Science handled by Subject Matter Experts and the external editor. The editor of this book selected which SN content should be auto-summarized and decided its order of appearance. Please be aware that the auto-summaries consist of original sentences, but are not representative of its original paper, since we do not show the full length of the publication. Please note that only published SN content is represented here, and that machine-generated books are still at an experimental stage.

Machine generated keywords: extremely, exciton, density, change, rate, carrier density, separation, metal halide, relaxation, mechanism, resistance, auger, surface, metal, perovskite perovskite

Molecularly Thin Two-Dimensional Hybrid Perovskites with Tunable Optoelectronic Properties Due to Reversible Surface Relaxation

DOI: https://doi.org/10.1038/s41563-018-0164-8

Originators: Leng, K., Abdelwahab, I., Verzhbitskiy, I. *et al.* Molecularly thin two-dimensional hybrid perovskites with tunable optoelectronic properties due to reversible surface relaxation. *Nature Mater* 17, 908–914 (2018). https://doi.org/10.1038/s41563-018-0164-8

Abstract-Summary

Due to their layered structure, two-dimensional Ruddlesden–Popper perovskites (RPPs), composed of multiple organic/inorganic quantum wells, can in principle be exfoliated down to few and single layers.

We have synthesized centimetre-sized, pure-phase single-crystal RPP perovskites $(CH_3(CH_2)_3NH_3)_2(CH_3NH_3)_{n-1}Pb_nI_{3n+1}$ (n = 1–4) from which single quantum well layers have been exfoliated.

A highly efficient photodetector was fabricated using a molecularly thin n = 4 RPP crystal, showing a photogain of 10^5 and an internal quantum efficiency of ~34%.

Main

Researchers in the last two decades have studied the RPP perovskites and its higher homologues to understand how their physical properties scale with the thickness of the quantum well; that is n = 4 crystal has a thicker quantum well than n = 1 (refs [1–7]).

A potentially rich area to explore is the molecular analogue of graphene in these layered RPP crystals when they are scaled to molecular thickness or single unit cell (that is, single quantum well) dimensions [8, 9].

With the aim of investigating whether the properties of 2D perovskites are unique at the molecularly thin limit, we synthesized centimetre-sized crystals grown by a temperature-programmed crystallization method [10], from which large-sized monolayer perovskite (defined here as mono-unit cell thickness) sheet can be exfoliated.

By studying the photoresponsivity of the single-crystal perovskites as a function of thickness, we further discover tunnelling of carriers or excitons across the hybrid interlayers, leading to efficient photocurrent generation.

Structural and optical properties of RPP single crystals

The optical properties of both bulk and monolayer flakes were investigated by photoluminescence and optical absorption measurements.

It is also obvious that the photoluminescence peaks of the exfoliated molecularly thin flakes are much sharper than those for bulk crystals.

Reversible order–disorder transition in RPP monolayers

We propose that the reversible spectral evolution is due to a dynamic equilibrium involving structural reorientation of the organic chains in the easily deformed 2D perovskite flakes, in which the hBN protection layer plays the important role of confining the organic cations and preventing their escape, thereby increasing their probability of returning to the lattice following a laser annealing process.

To obtain direct visual evidence of structural relaxation, we used a Q-plus non-contact atomic force microscope (nc-AFM) to probe the relaxation of the organic chains when the exfoliated perovskite flake was subjected to thermal stress, simulating the thermal annealing effects by laser irradiation.

Insights into the changes in the DOS of the perovskite monolayer as a result of surface relaxation and the formation of organic cation vacancies can be obtained by performing first-principles calculations with density functional theory (DFT), utilizing the Vienna ab initio simulation package (VASP) [11].

Photodetectors based on monolayer and bulk RPP flakes

In the dynamically disordered energy landscape induced by laser annealing, another effect that can be convoluted with the presence of trap states is the self-doping effect due to the generation of organic cation vacancies.

Simultaneous photoluminescence measurement was used to distinguish derelaxed state (~650 nm photoluminescence) from relaxed state (~675 nm photoluminescence) of the $n = 4$ perovskite photodetector.

We attribute the high photocurrent yield to the high photoconductive gain of the perovskite device due to the presence of shallow trap states as well as the self-doping effect induced by organic cation vacancies [12].

We observed that increased laser irradiation time or power of the unprotected perovskite sample will lead to a higher defect concentration and thus create more recombination centres (hence shorter carrier lifetime), but the situation is different in the case of hBN-protected flakes due to the ability of the defects to be repaired by further laser annealing, thus limiting the concentration of defects in the relaxed state.

Outlook

Molecularly thin perovskites display optical properties that are different from their bulk counterparts due to their easily deformable nature and surface relaxation effects.

Confining the surface of ultrathin (<6 nm thickness) RPP perovskites with hBN allows dynamic surface relaxation to be activated by thermal energy without introducing permanent defects.

One way to exploit the surface relaxation effect is through interface engineering of these 2D perovskites and other 2D materials, whereby strain-tuning of coupling at the interface can lead to advanced hybrid materials with novel properties.

Methods

The size of the focused laser spot was ~1 μm^2, and the maximum responsivities of 11.3, 25 and 71.9 mA W^{-1} were achieved at 36 nW incident power for a monolayer in a relaxed state, bilayer and bulk flakes, with photocurrents of 0.41, 0.87 and 2.6 nA, respectively.

Based on a defocused laser spot of ~4 mm and an active working area of 4 μm^2 for our device, the light power received by the device is calculated to be 6.36×10^{-6} nW at 20 nW incident power [13].

The laser power (P_{abs}) received by our bulk device was 6.36×10^{-6} nW at 20 nW incident power, with I_{ph} of 4.73×10^{-10} A under this power, so the calculated G was 1.73×10^5.

Time-dependent photoluminescence measurements were conducted using a 532 nm continuum-wavelength laser focused by a ×50 microscope objective lens (NT-MDT), with a neutral density filter to tune the laser power density.

Design Principles for Electronic Charge Transport in Solution-Processed Vertically Stacked 2D Perovskite Quantum Wells

DOI: https://doi.org/10.1038/s41467-018-04430-2

This article is licensed under a Creative Commons Attribution 4.0 International License, which permits use, sharing, adaptation, distribution and reproduction in any medium or format, as long as you give appropriate credit to the original author(s) and the source, provide a link to the Creative Commons license, and indicate if changes were made. The images or other third party material in this article are included in the article's Creative Commons license, unless indicated otherwise in a credit line to the material. If material is not included in the article's Creative Commons license and your intended use is not permitted by statutory regulation or exceeds the permitted use, you will need to obtain permission directly from the copyright holder. To view a copy of this license, visit http://creativecommons.org/licenses/by/4.0/.

Originators: Tsai, H., Asadpour, R., Blancon, JC. *et al.* Design principles for electronic charge transport in solution-processed vertically stacked 2D perovskite quantum wells. *Nat Commun* 9, 2130 (2018). https://doi.org/10.1038/s41467-018-04430-2

Abstract-Summary

State-of-the-art quantum-well-based devices such as photovoltaics, photodetectors, and light-emission devices are enabled by understanding the nature and the exact mechanism of electronic charge transport.

Further improvements will require an understanding of the charge transport mechanisms, which are currently unknown and further complicated by the presence of strongly bound excitons.

These findings establish the fundamental guidelines for the molecular and device design for layered 2D perovskite-based photovoltaics and optoelectronic devices, and are relevant for other similar quantum-confined systems.

Introduction

To realize the tremendous potential of layered 2D perovskites for photovoltaics and other efficient optoelectronic devices, it is critical to understand the fundamental physical mechanisms that limit the efficient charge transport in these systems.

Through extensive device characterization and modeling, we elucidate the dominant charge transport mechanism during solar cell operation and identify the key bottlenecks that limit the overall efficiency in layered 2D perovskites.

Our model self-consistently reproduces our experimentally observed device behavior, thus validating that the key bottleneck limiting the photocurrent collection is indeed field-dependent charge separation through the barriers.

Our results provide the fundamental guidelines for the design of layered 2D perovskites for high-efficiency photovoltaic devices, which will require improving the light absorption, engineering highly doped contacts to facilitate efficient charge separation and collection for thicker films, or investigating the incorporation of conducting (organic or inorganic) spacer molecules to reduce the potential barriers.

Results

The J_{SC} for 3D perovskite cell is less sensitive to the film thickness in the 300–500 nm range because optical excitation produces free carriers [14–16] that can be efficiently collected at the contacts, even for thicker layers as the cell operates in the non-Langevin type of recombination regime.

The photogenerated carrier collection, especially at the center of the film for the layered perovskite system, is strongly field dependent, leading to a reduction in the J–V slope near OC, while the SC slope remains flat.

The free carriers can be collected near SC quite efficiently, as evidenced by the linear relationship of J_{SC} versus power curve within this thickness range, and the reduction in J_{SC} from 200 to 375 nm is thus attributed to monomolecular loss.

The photogenerated carriers can only undergo a partial separation, and non-separated carriers generated in the center of the film recombine strongly through a monomolecular process, which results in strong field dependence near SC as well as OC.

Discussion

And discussions, we establish that the key bottleneck in the layered perovskite quantum well photovoltaic devices is the field-limited carrier collection.

Independent of whether the photogenerated carrier is an exciton or free carrier, a strong electric field is necessary for the carriers to surmount the potential barrier before they recombine.

Based on our results, we have identified the key bottleneck for charge transport for vertically stacked layered perovskite quantum well photovoltaic devices arising from recombination losses of charge carriers across potential barriers created.

These findings present opportunities for improved design of 2D perovskite structures, where long-range vertical packing for facilitating conducting pathways can be achieved that results in reduced numbers of potential barriers.

Methods

Indium tin oxide-coated glass slides were cleaned through standard steps by ultrasonication bath with distilled water, acetone, and isopropyl alcohol for 30 min, respectively.

After 5000 rpm for 45 s, spin coating the PEDOT:PSS layer formed in thickness of around 50 nm serves as HTL.

After drying at 130 °C for 30 min, all the substrates were transferred into argon-filled glovebox for perovskite layer deposition.

We followed our previously developed fabrication procedure [6, 17–21], first preheat the substrate at 110 °C for 5 min and then quickly transfer to the spin-coating chunk; 100 µl of previously prepared precursors was then dropped onto the hot substrate and the spin coater started immediately to prevent temperature from quenching.

The ETL was done by spin coating PCBM (Nano-C with purity over 99.5%) solution at 1000 rpm for 45 s. Finally, the coated substrates were mounted into vacuum chamber inside the glovebox for aluminum deposition.

Extremely Reduced Dielectric Confinement in Two-Dimensional Hybrid Perovskites with Large Polar Organics

DOI: https://doi.org/10.1038/s42005-018-0082-8

This article is licensed under a Creative Commons Attribution 4.0 International License, which permits use, sharing, adaptation, distribution and reproduction in any medium or format, as long as you give appropriate credit to the original author(s) and the source, provide a link to the Creative Commons license, and indicate if changes were made. The images or other third party material in this article are included in the article's Creative Commons license, unless indicated otherwise in a credit line to the material. If material is not included in the article's Creative Commons license and your intended use is not permitted by statutory regulation or exceeds the permitted use, you will need to obtain permission directly from the copyright holder. To view a copy of this license, visit http://creativecommons.org/licenses/by/4.0/.

Originators: Cheng, B., Li, TY., Maity, P. *et al.* Extremely reduced dielectric confinement in two-dimensional hybrid perovskites with large polar organics. *Commun Phys* 1, 80 (2018). https://doi.org/10.1038/s42005-018-0082-8

Abstract-Summary

We theoretically predict that an extremely low exciton binding energy can be achieved in 2D perovskites by using high dielectric-constant organic components.

We demonstrate that in $(HOCH_2CH_2NH_3)_2PbI_4$, whose organic material has a high dielectric constant of 37, the dielectric confinement is largely reduced, and the exciton binding energy is 20-times smaller than that in conventional 2D perovskites.

Such low dielectric-confined 2D perovskites show excellent carrier extraction efficiency, and outstanding humidity resistance compared to conventional 2D perovskites.

Introduction

Since the conventional organic linkers used in 2D perovskites, such as $C_6H_5(CH_2)_2NH_3^+$ and $C_4H_9NH_3^+$, have small dielectric constants, dielectric confinement also emerges and makes the exciton-binding energy even larger [22–25].

The strong combined effect of quantum and dielectric confinement makes the carrier mobility in 2D perovskites approximately two orders of magnitude lower than that in 3D perovskites [26].

We hypothesize that dielectric confinement in 2D perovskites can be largely reduced by using a high-dielectric constant organic component.

We demonstrate that $(HOCH_2CH_2NH_3)_2PbI_4$ (2D_EA perovskites) [27], in which the dielectric constant of the organic component is as high as 37.7 [28], has a 20-times smaller exciton- binding energy and dozens of times larger carrier mobility compared to high-dielectric confinement 2D perovskites.

We find that low-dielectric-confined 2D perovskites are highly stable under humidity, due to the strong hydrogen bonding between the organic layers.

Result

Temperature-dependent PL measurements were performed on both 2D_EA and 2D_PEA perovskite single crystals to determine the exciton-binding energies.

The slow GSB recovery and excited state decay results are consistent with our exciton-binding energy findings, and strongly prove that the 2D_EA perovskites have a longer carrier lifetime and more efficient exciton dissociation than the 2D_PEA perovskites.

Due to the low-binding energy and high carrier mobility, the 2D_EA perovskites are expected to present much better photo-excited carrier extraction efficiency compared to conventional 2D perovskites in terms of photoresponsivity, photo gain, and response time.

Based on the results of photoresponsivity, photo gain, and response time we conclude that due to the low binding energy and high carrier mobility, the 2D_EA perovskite show signficantly improved carrier extraction efficiency, which could have a huge impact on the optoelectronic device performance, particularly for solar cells and photodetectors.

Discussion

We have investigated the effect of the organic layer with different dielectric constants on the exciton-binding energy and carrier mobility in 2D perovskites.

We have shown that in the low dielectric-confined 2D perovskites the exciton-binding energy is ~13 meV, which is 20 times smaller than that in the high dielectric-confined 2D perovskites.

In the low dielectric-confined 2D perovskite single crystal we demonstrated excellent photo-excited carrier extraction efficiency due to the largely improved carrier mobility and reduced exciton-binding energy, and outstanding moisture resistance due to the strong hydrogen bonding between the organic layers.

Method

Two measurements were performed to collect the absorption spectra for the 2D perovskite nanosheets.

In the first measurement, the absolute values of reflection and transmission at a wavelength of 514.5 nm (Ar ion laser) were measured for the extraction of the absorption value using a lock-in amplifier coupled with a silicon photodiode.

For transmission measurement, the transmitted light was collected using a ×20 objective.

The probe pulses (UV–visible and NIR wavelength continuum, white light) were generated by passing another fraction of the 800 nm pulses through a 2-mm-thick calcium fluoride (CaF_2) crystal.

To confirm the size and uniform intensity distribution of the incident light spot, the optical profile was measured by a beam profiling camera (Ophir Spiricon silicon CCD camera SP620U) and the power at each wavelength was measured using a calibrated photodiode (Newport 818-UV).

Layer-Edge Device of Two-Dimensional Hybrid Perovskites

DOI: https://doi.org/10.1038/s41467-018-07656-2

This article is licensed under a Creative Commons Attribution 4.0 International License, which permits use, sharing, adaptation, distribution and reproduction in any medium or format, as long as you give appropriate credit to the original author(s) and the source, provide a link to the Creative Commons license, and indicate if changes were made. The images or other third party material in this article are included in the article's Creative Commons license, unless indicated otherwise in a credit line to the material. If material is not included in the article's Creative Commons license and your intended use is not permitted by statutory regulation or exceeds the permitted use, you will need to obtain permission directly from the copyright holder. To view a copy of this license, visit http://creativecommons.org/licenses/by/4.0/.

Originators: Cheng, B., Li, TY., Wei, PC. *et al.* Layer-edge device of two-dimensional hybrid perovskites. *Nat Commun* 9, 5196 (2018). https://doi.org/10.1038/s41467-018-07656-2

Abstract-Summary

Two dimensional layered organic-inorganic hybrid perovskites (2D perovskites) are potential candidates for next generation photovoltaic device.

The out-of-plane surface perpendicular to the superlattice plane of 2D perovskites (layer-edge surface) has presented several exotic behaviors, such as layer-edge states which are found to be crucial for improving the efficiency of 2D perovskite solar cells.

We observe the electronic and opto-electronic behavior in layer-edge device of 2D perovskites.

Introduction

In the two-dimensional-layered-organic–inorganic hybrid perovskites (2D perovskites), the out-of-plane surface perpendicular to the superlattice plane, which is named as layer-edge surface, has unique organic–inorganic staggered superlattice pattern, as a straightforward result of multiple quantum well structures.

Those impressive results presented fancy properties of the layer-edge surface in 2D perovskites, but several other important fundamental researches are still required, especially the electronic transport properties which is one of the key factors to determine the performance of opto-electronic devices.

We succeed to grow centimeter-wide and millimeter-thick single crystalline 2D perovskites, and fabricate the metal–semiconductor-metal devices on the layer-edge surface in the direction parallel to the Pb–I plane (defined as layer-edge 0° device) and perpendicular to the Pb–I plane (defined as layer-edge 90° device), respectively.

Transport measurements are performed in these layer-edge devices of 2D perovskites, showing highly anisotropic electrical and opto-electronic conductance.

Results

To study the photo-response dynamics in layer-edge surface devices, we measured the time-dependent photo current of 2D perovskites ($n = 1, 2, 3$) and 3D perovskites.

The result in 2D perovskites is expected to be quite different because of its layered structure and different organic groups adopted, especially in the layer-edge device some abnormal humidity sensitivity may show up.

The 2D perovskite layer-edge 0° device with $n = 1$ presents the best humidity sensitivity, and relative outstanding response time.

Compared to the best humidity sensor based on hybrid perovskite so far [29], the performance of humidity sensitivity and response time in 2D perovskite layer-edge device with $n = 1$ are much more excellent.

As n increases, the humidity sensitivity of layer-edge device is decreased.

Discussion

We succeed to fabricate the layer-edge surface device of 2D hybrid perovskites with different inorganic layer thickness, and investigate their transport properties.

The existence of large anisotropy of electronic transport in 2D perovskites indicates the importance of the choice of crystalline orientation in the layer-edge device.

The humidity sensing properties of layer-edge device in 2D perovskite strongly depend on the organic groups adopted, indicating urgent requirement of research on 2D perovskites with different organic layers in the future.

Our study has revealed the unusual transport properties of layer-edge surface, and indicated potential applications of humidity sensor based on 2D perovskite single crystals.

Methods

2D perovskites ($HOC_2H_4NH_3$) single crystals are grown using a cooling method.

In order to grow n = 2 perovskites, we prepare 16 ml HI (57 wt. %) and put in 5.5 g PbI_2 powder.

For n = 3 perovskite single crystal, we prepare 12 ml HI solution, we need to add 5 g PbI_2 powder, 0.4 ml MA and 0.8 ml EA.

The measured resistance between electrodes 2 to 3 include the intrinsic resistance of crystal (R_i) and contact resistance (R_c) between the crystal and the electrodes, i.e., $R_i + R_c = V_b/I_{2\text{–}3}$.

We can exclude the contact resistance and only measure the intrinsic resistance between electrode 2 to 3: $R_i = V_{2\text{–}3}/I_{1\text{–}4}$.

In order to eliminate the effect of different channel geometry when we calculate the anisotropy of resistance, we do some corrections:Where I is the current, and l is the channel width.

Long-Range Exciton Transport and Slow Annihilation in Two-Dimensional Hybrid Perovskites

DOI: https://doi.org/10.1038/s41467-020-14403-z

This article is licensed under a Creative Commons Attribution 4.0 International License, which permits use, sharing, adaptation, distribution and reproduction in any medium or format, as long as you give appropriate credit to the original author(s) and the source, provide a link to the Creative Commons license, and indicate if changes were made. The images or other third party material in this article are included in the article's Creative Commons license, unless indicated otherwise in a credit line to the material. If material is not included in the article's Creative Commons license and your intended use is not permitted by statutory regulation or exceeds the permitted use, you will need to obtain permission directly from the copyright holder. To view a copy of this license, visit http://creativecommons.org/licenses/by/4.0/.

Originators: Deng, S., Shi, E., Yuan, L. *et al.* Long-range exciton transport and slow annihilation in two-dimensional hybrid perovskites. *Nat Commun* 11, 664 (2020). https://doi.org/10.1038/s41467-020-14403-z

Abstract-Summary

Exciton transport and annihilation are two key processes in determining device efficiencies; however, a thorough understanding of these processes is hindered by that annihilation rates are often convoluted with exciton diffusion constants.

We employ transient absorption microscopy to disentangle quantum-well-thickness-dependent exciton diffusion and annihilation in two-dimensional perovskites, unraveling the key role of electron-hole interactions and dielectric screening.

The combination of long-range exciton transport and slow annihilation highlights the unique attributes of two-dimensional perovskites as an exciting class of optoelectronic materials.

Introduction

The structure of 2D perovskites is extremely programmable; for instance, continuous tuning of the exciton emission over a wide range of energy can be achieved by either changing n or the halide in the composition [4, 30].

To harvest excitons in a device, a thorough understanding of exciton transport and annihilation in 2D perovskites is necessary.

Despite the impressive progress made in the optoelectronic devices based on 2D perovskites [30], the understanding of exciton transport is extremely limited and currently there is no direct measurement on exciton diffusion.

To address this challenge, here we employ transient absorption microscopy (TAM) as a direct means to image exciton population in space and in time to disentangle exciton diffusion and annihilation in 2D perovskites.

These results showcase the unique ability of 2D perovskites in achieving a combination of large exciton binding energy, long-range exciton transport, and slow annihilation, suggesting their large potential in optoelectronic applications.

Results

Further, third-order Auger recombination of free charge carriers is also possible for $n > 1$ quantum wells, as a finite free carrier population exists due to smaller exciton binding energy [31].

The exciton diffusion constants measured for the 2D perovskites are consistent with the carrier diffusion constant of around 1 $cm^2\ s^{-1}$ at the 3D limit of $n = \infty$ [32].

It is possible that photon recycling that has been proposed to enhance carrier diffusion in 3D perovskites contributes to the long-range exciton transport observed here [33–35].

For all five structures, exciton diffusion constants are significantly larger than annihilation rates ($D \gg k_2$).

The observation of enhanced EEA rates at larger exciton binding energies in 2D perovskites are consistent with those reported in single-walled carbon nanotubes and colloidal quantum dots [36–39].

For solar cells, quantum wells with large n are suitable because of the combined benefits of the small exciton binding energy and long-range diffusion.

Discussion

A major challenge in realizing optoelectronic applications for low-dimensional nanostructures is to achieve long-range transport and suppressed annihilation while maintaining large exciton binding energy.

The exciton binding energies of 2D perovskites are comparable to those found in TMDCs but their annihilation rates are orders of magnitude lower.

Further, 2D perovskites are also promising as solar cell materials, with long-range exciton diffusion over hundreds of nanometers.

Methods

For the micro-absorption measurements, a halogen tungsten lamp was used as white light source, which was focused by objective on sample to form light spot with size of around 2 μm.

The reflective or transmitted light was collected by an objective (×40, NA = 0.6, Nikon), and detected by a spectrometer (Andor Shamrock 303i) and CCD (Andor Newton 920) for spectral analysis.

For the PL measurements, a picosecond pulse laser with wavelength of 447 nm was used as excitation light and an objective (×40, NA = 0.6, Nikon) was employed to focus the excitation light on the sample and collect the epic scattered PL light.

The output from two independent optical parametric amplifiers (OPAs, 400 kHz, ~300 fs, Light Conversion) served as pump and probe beam respectively, which were focused on sample with an objective (×60, NA = 0.95, Nikon).

Overcoming the Exciton Binding Energy in Two-Dimensional Perovskite Nanoplatelets by Attachment of Conjugated Organic Chromophores

DOI: https://doi.org/10.1038/s41467-020-15869-7

This article is licensed under a Creative Commons Attribution 4.0 International License, which permits use, sharing, adaptation, distribution and reproduction in any medium or format, as long as you give appropriate credit to the original author(s) and the source, provide a link to the Creative Commons license, and indicate if changes were made. The images or other third party material in this article are included in the article's Creative Commons license, unless indicated otherwise in a credit line to the material. If material is not included in the article's Creative Commons license and your intended use is not permitted by statutory regulation or exceeds the permitted use, you will need to obtain permission directly from the copyright holder. To view a copy of this license, visit http://creativecommons.org/licenses/by/4.0/.

Originators: Gélvez-Rueda, M.C., Fridriksson, M.B., Dubey, R.K. *et al.* Overcoming the exciton binding energy in two-dimensional perovskite nanoplatelets by attachment of conjugated organic chromophores. *Nat Commun* 11, 1901 (2020). https://doi.org/10.1038/s41467-020-15869-7

Abstract-Summary

We demonstrate a novel approach to achieve efficient charge separation in dimensionally and dielectrically confined two-dimensional perovskite materials.

Two-dimensional perovskites generally exhibit large exciton binding energies that limit their application in optoelectronic devices that require charge separation such as solar cells, photo-detectors and in photo-catalysis.

We show that by incorporating a strongly electron accepting moiety, perylene diimide organic chromophores, on the surface of the two-dimensional perovskite nanoplatelets it is possible to achieve efficient formation of mobile free charge carriers.

Introduction

2D hybrid perovskites differ from three-dimensional perovskites, which are currently among the most promising materials for solar cell applications, due to the presence of large organic cations that lead to the formation of 2D layers of inorganic metal-halide octahedra separated by the organic cations.

Endless possibilities to engineer the electronic structure of 2D hybrid perovskites by introduction of functional organic molecules, for instance strong electron donors or acceptors.

In some recent studies, functional organic cations such as conjugated molecules [40, 41] and charge-transfer complexes [42, 43] have been introduced in 2D hybrid perovskites.

We explore the introduction of functional organic cations in 2D perovskites in order to enhance their functionality by, for example, inducing change separation from the inorganic octahedral layer into the organic chromophores.

Results

We are convinced that the most direct proof of the attachment of the PDIs to the perovskite NPLs is shown by our TA measurements photoexciting the PDI (510 nm excitation) as described below in this work.

To determine whether the deactivation of the excitons in the NPLs is indeed due to electron transfer from the $CsPbBr_3$ NPLs to PDI, we have studied the exciton dynamics by femtosecond TA measurements.

In order to determine whether the inverse process, HT from the PDI molecules to the $CsPbBr_3$ NPLs also takes place, we have studied the exciton dynamics by TA measurements photoexciting at 510 nm.

The TA measurements presented above clearly show that charge carriers with lifetimes extending into the nanosecond regime are formed, either on exciting the NPLs or the attached PDI molecules.

These measurements unequivocally prove that the charge separation is efficient and mobile-free charge carriers (free holes) are formed when the PDI molecules are attached to the NPLs.

Discussion

We have demonstrated how improved functionality, in this case efficient charge carrier separation and long-range free carrier conduction, can be achieved in 2D perovskite NPLs by smart synthetic design attaching strong electron acceptors: amine-functionalized PDI molecules.

By the unique combination of several ultrafast spectroscopy techniques, such as TA, picosecond PL spectroscopy and microwave conductivity measurements, we have shown that photoexcitation of either the NPLs or the PDIs leads to charge separation, and we have extracted the charge-transfer rates for electrons and holes ($k_{ET} = 0.079\ ps^{-1}$ and $k_{HT} = 0.65\ ps^{-1}$).

This is the first time that improved charge separation has been directly shown in 2D perovskites by incorporation of a functional organic group in the same material (attached to the inorganic octahedral layer and not as heterojunction hampered by morphological issues).

Methods

To synthesize $CsPbBr_3$ NPLs with 4 monolayers: 150 μL Cs-oleate precursor (prepared by dissolving 0.1 mmol Cs_2CO_3 in 10 mL oleic acid at 100 °C) was added to 1.2 mL $PbBr_2$ precursor (prepared by dissolving 0.1 mmol $PbBr_2$ in 10 mL toluene in the presence of 100 μL oleic acid and oleylamine at 100 °C) under continuous stirring.

Samples for optical measurements in solution were prepared by diluting the stock solution of washed $CsPbBr_3$ NPLs with hexane.

The TRMC technique measures the change in microwave (8–9 GHz) power after pulsed excitation (repetition rate 10 Hz) at 400 and 510 nm in a temperature range of 295–93 K. The time resolution is limited by the width of the laser pulse (3.5 ns fwhm) and the response time of the system (18 ns).

Reversible Multicolor Chromism in Layered Formamidinium Metal Halide Perovskites

DOI: https://doi.org/10.1038/s41467-020-19009-z

This article is licensed under a Creative Commons Attribution 4.0 International License, which permits use, sharing, adaptation, distribution and reproduction in any medium or format, as long as you give appropriate credit to the original author(s) and the source, provide a link to the Creative Commons license, and indicate if changes were made. The images or other third party material in this article are included in the article's Creative Commons license, unless indicated otherwise in a credit line to the material. If material is not included in the article's Creative Commons license and your intended use is not permitted by statutory regulation or exceeds the permitted use, you will need to obtain permission directly from the copyright holder. To view a copy of this license, visit http://creativecommons.org/licenses/by/4.0/.

Originators: Rosales, B.A., Mundt, L.E., Allen, T.G. *et al.* Reversible multicolor chromism in layered formamidinium metal halide perovskites. *Nat Commun* 11, 5234 (2020). https://doi.org/10.1038/s41467-020-19009-z

Abstract-Summary

Metal halide perovskites feature crystalline-like electronic band structures and liquid-like physical properties.

We leverage the low formation energy of metal halide perovskites to demonstrate multicolor reversible chromism.

We synthesized layered Ruddlesden-Popper $FA_{n+1}Pb_nX_{3n+1}$ (FA = formamidinium, X = I, Br; n = number of layers = 1, 2, 3 … ∞) and reversibly tune the dimensionality (n) by modulating the strength and number of H-bonds in the system.

Each colored phase exhibits distinct optoelectronic properties characteristic of 2D superlattice materials with tunable quantum well thickness.

Introduction

2D MHPs are classified as Ruddlesden-Popper [44] or Dion-Jacobson [45, 46] phases of the general formula $A'_2A_{n-1}M_nX_{3n+1}$ or $A'A_{n-1}M_nX_{3n+1}$, respectively, where A′ = 1+ or 2+ cation, A = 1+ cation, and n = 1, 2, 3, … ∞ is the number of connected 2D octahedra layers.

There are two recognized mechanisms for reversible chromism in MHP materials: (1) crystal phase transformation [47] and (2) molecular intercalation [48].

MHPs are inherently thermochromic materials exhibiting significant optical changes induced by crystal phase transitions between the α-phase, a black high-symmetry perovskite phase composed of corner-sharing $[MX_6]^{4-}$ octahedra, and the δ-phase, a yellow-to-colorless non-perovskite hexagonal or orthorhombic phase composed of face- or edge-sharing $[MX_6]^{4-}$ octahedra, respectively [49–51].

We leverage their liquid-like physical properties to demonstrate a third mechanism of dynamic chromism in MHPs through reversible layer formation and coalescence to form compounds that span 2D FA_2PbX_4 (n = 1) to 3D α-$FAPbX_3$ (n = ∞) and finally 1D δ-$FAPbI_3$.

Results

When assembled properly, composite films will reversibly change color rapidly (<1–20 min depending on solvent vapor flow rate) and over many cycles (>10).

Reversible chromism of composite $FA_{n+1}Pb_nX_{3n+1}$ films is strongly dependent on the affinity of the solvent vapor to form H-bonds with the FAX reservoir and $FA_{n+1}Pb_nX_{3n+1}$ constituents.

Before a change in color occurs (RH < 70%), the increased intensity of the O–H stretching mode located between 3550 and 3200 cm^{-1} suggests increasing amounts of H_2O are incorporated into the film and are intermolecularly H-bonded with other H_2O molecules or with FAI in the reservoir [52, 53].

We believe electron and hole mobilities (represented by $\Sigma\mu$) increase concomitantly with charge yield [54] and argue the $\phi\Sigma\mu$ values of the orange phase are closer to the brown phase likely due to an appreciable density of $n > 4$ domains that dominate the fp-TRMC signal over smaller (n = 1–3) layers.

Discussion

We synthesized layered FA-based MHPs of the general formula $FA_{n+1}Pb_nX_{3n+1}$ (X = I, Br) and their mixed-halide compositions and show reversible multicolor chromism based on two different stimuli: solvent vapor and temperature.

The interactions between each species results in a structural equilibrium between 2D $FA_{n+1}Pb_nX_{3n+1}$ layers, 3D α-$FAPbX_3$ domains, and 1D δ-$FAPbI_3$ domains that can be controlled through modulating the strength and number of H-bonds between the reservoir, MHP, and solvent vapor.

Our work will enable a new generation of functional materials that couple tunable and reversible chromism with the extraordinary optoelectronic properties of MHP materials.

Methods

Optical measurements were collected on films prepared with a 326 ± 22 nm thick Al_2O_3 NP scaffold.

The samples were measured at an incident angle of 3 degrees and an incident X-ray wavelength of 0.9744 Å. A Rayonix MX225 2D detector was used to collect data, and a LaB_6 standard used to calibrate the data.

All samples were excited at 450 nm with fluences in the range of ca 1×10^{11} to 2×10^{15} photons cm^{-2} depending on the strength of the sample response and phase stability during illumination.

Maintaining the RH levels necessary to stabilize the brown phase during measurements was achieved by sealing samples in a leak-resistant microwave cavity within the humidified glovebox described previously.

A 500 nm long pass filter was used to filter out excitation light (450 nm) and photoluminescence spectra were typically acquired by averaging 60 seconds to monitor phase stability on a minute-to-minute basis during measurements.

Reducing the Impact of Auger Recombination in Quasi-2D Perovskite Light-Emitting Diodes

DOI: https://doi.org/10.1038/s41467-020-20555-9

This article is licensed under a Creative Commons Attribution 4.0 International License, which permits use, sharing, adaptation, distribution and reproduction in any medium or format, as long as you give appropriate credit to the original author(s) and the source, provide a link to the Creative Commons license, and indicate if changes were made. The images or other third party material in this article are included in the article's Creative Commons license, unless indicated otherwise in a credit line to the material. If material is not included in the article's Creative Commons license and your intended use is not permitted by statutory regulation or exceeds the permitted use, you will need to obtain permission directly from the copyright holder. To view a copy of this license, visit http://creativecommons.org/licenses/by/4.0/.

Originators: Jiang, Y., Cui, M., Li, S. *et al.* Reducing the impact of Auger recombination in quasi-2D perovskite light-emitting diodes. *Nat Commun* 12, 336 (2021). https://doi.org/10.1038/s41467-020-20555-9

Abstract-Summary

Rapid Auger recombination represents an important challenge faced by quasi-2D perovskites, which induces resulting perovskite light-emitting diodes' (PeLEDs) efficiency roll-off.

Auger recombination rate is proportional to materials' exciton binding energy (E_b).

Auger recombination can be suppressed by reducing the corresponding materials' E_b.

Devices exhibit a record luminance of 82,480 cd m^{-2} due to the suppressed efficiency roll-off, which represent one of the brightest visible PeLEDs yet.

Introduction

In quasi-2D films, the amplified carrier density is created at the recombination center due to the efficient energy transfer.

The threshold is quite low for Auger recombination to become dominated in quasi-2D perovskites.

Auger recombination rate is proportional to the third power of the E_b in strongly confined 1D material [36].

Quasi-2D perovskites should exhibit fast Auger recombination because of their large E_b.

Structure engineering innovation has yet been tentatively explored in quasi-2D perovskites to tailor Auger recombination.

Auger recombination rate is proportional to the third power of E_b, in strongly confined 1D material [36].

Reducing E_b can suppress Auger recombination in 1D material.

We conceive to explore whether reducing E_b could also alter the Auger recombination rates in quasi-2D perovskites.

The innovation here paves the way for regulating Auger recombination of quasi-2D perovskite optoelectronics.

Results

In specific, I_{PL0} is linear with excitation density under low fluence (~10^{16} cm^{-3} for PEA film and ~4×10^{16} cm^{-3} for p-FPEA film, respectively), indicating the characteristics of exciton recombination (monomolecular process).

We did find the higher threshold for PLQY declining in p-FPEA film, illustrating the Auger recombination constant (k_3) decreased.

It is worth mentioning that the trimolecular Auger recombination constants are more than one-order-of magnitude lower for p-FPEA case when compared to the PEA analog, reconfirming that the Auger recombination rate has been notably suppressed through p-FPEA substitution, and in good agreement with the above TRPL analysis.

The fitted first-order recombination rates are determined to be 1.1×10^7 s^{-1} and 5.3×10^6 s^{-1} for CF_3KO_3S-PEA and CF_3KO_3S-p-FPEA films, respectively.

The PLQY of CF_3KO_3S-p-FPEA quasi-2D film displays nearly constant value within a broad region of carrier density, demonstrating the first-order exciton recombination and bimolecular recombination always control over the recombination channel.

Discussion

Rapid nonradiative Auger recombination rate is an important challenge faced by quasi-2D perovskites, which causes significant efficiency roll-off and impedes their further commercialization.

Auger recombination rate is proportional to the materials' exciton binding energy (E_b); thereby, Auger process might be suppressed by reducing the corresponding E_b in principle.

Molecular passivation agent, CF_3KO_3S, is used to reduce trap-assistant recombination rate, which enables the film's PLQY to become independent with exciton recombination rate decline.

Suppressed Auger recombination also reduces the resulting Joule heating, thus notably elongates device stability in particular under high current density.

Methods

Crystallographic data of p-$FPEAPbBr_4$ single crystal was collected on a Bruker APEX II CCD diffractometer with Mo Kα radiation (50 kV, 30 mA), followed by integrating and scaling with multi-scan absorption correction using SAINT and SADABS.

The excitation source was a picosecond pulsed laser with a pulse width of below 100 ps and a repetition rate of 800 kHz at 355 nm.

The pulse beam was excited by a Ti:sapphire regenerative amplifier (Legend Elite-1K-HE; pulse width, 35 fs; pulse energy, 7 mJ per pulse; repetition rate, 1 kHz; 800 nm), and seeded with a mode-locked Ti:sapphire laser system (Mica 5) and an Nd:YLF laser (EvolutIon 30) pumped.

The larger portion was passed an optical parametric amplifier (TOPAS-800-fs, Coherent) to generated the pump pulse (pulse width, around 100 fs; pulse energy, 18.8 μJ per pulse; repetition rate, 1 kHz; 365 nm).

Electrical Doping in Halide Perovskites

DOI: https://doi.org/10.1038/s41578-021-00286-z

Originators: Euvrard, J., Yan, Y. & Mitzi, D.B. Electrical doping in halide perovskites. *Nat Rev Mater* 6, 531–549 (2021). https://doi.org/10.1038/s41578-021-00286-z

Abstract-Summary

Electrical doping (that is, intentional engineering of carrier density) underlies most energy-related and optoelectronic semiconductor technologies.

For the intensely studied halide perovskite family of semiconductors, reliable doping remains challenging, owing to, for example, compensation from and facile migration of intrinsic defects.

In this Review, we first discuss the underlying fundamentals of semiconductor doping and then investigate different doping strategies in halide perovskites, including intrinsic defect, extrinsic defect and charge transfer doping, from an experimental as well as a theoretical perspective.

Introduction

The term 'doping' is broadly used in HaP literature to describe the introduction of any kind of impurity into the lattice to control crystal growth and stability or to tune optoelectronic properties [55, 56].

In this Review, we focus on electrical doping, that is, the introduction of an impurity or defect to generate free carriers or to shift the Fermi level.

Considerations of HaP doping have been explored early on with the observation that Sn-based perovskites have p-type semimetallic properties and can be tuned upon Sn oxidation [1, 57–61].

Efficient and reliable tuning of carrier density remains particularly challenging in HaPs, because major difficulties have arisen, including compensation from intrinsic defects.

Different strategies have been explored for HaP doping, which can be categorized into intrinsic defect doping, extrinsic defect doping and charge transfer doping.

Charge transfer doping occurs through exchange of carriers between the perovskite semiconductor and an external entity, such as a molecule or interface.

Fundamentals of electrical doping

Electrical doping of a semiconductor implies the addition of extra electrons (or holes) in the conduction (or valence) band to generate additional mobile carriers.

In intrinsic semiconductors, the electron n and hole p densities balance [62]: where N_C and N_V are the effective densities of states in the conduction and valence bands, respectively (dependent on effective masses and temperature), E_G is the semiconductor bandgap, k_B is the Boltzmann constant and T the temperature.

Although usually used to achieve higher electron or hole densities, doping can also be applied to compensate carrier concentrations (that is, to reduce carrier density or change the majority carrier type).

The space charge region extends exclusively in the semiconductor layer and the depletion width w of the space charge region decreases with increasing donor N_D and acceptor N_A densities for n-doped and p-doped semiconductors, respectively [63]:where ε_r is the relative permittivity of the semiconductor and ε_0 the vacuum permittivity.

Doping in halide perovskites

Close to intrinsic carrier densities of $\sim 10^7$ cm^{-3} for both p-type (Pb-poor) and n-type (Pb-rich) suggest, pending further confirming studies, a limited doping range achievable through precursor ratio variations for X=Br.

Extrinsic dopants cannot achieve n-type doping beyond what can be realized with intrinsic defects, highlighting the importance of charge compensation in HaPs and that careful control of growth conditions may favour more efficient extrinsic doping.

Although substitution of Sn^{2+} with a higher oxidation state, such as Sn^{4+}, should lead to n-type doping, p-doping has been experimentally observed, owing to the concomitant formation of Sn vacancies (V_{Sn}), which possess a low formation energy in Sn-based perovskites [64–66].

The acceptor molecule 2,3,5,6-tetrafluoro-7,7,8,8-tetracyanoquinodimethane (F4TCNQ) has been introduced in the precursor solution to p-dope $MAPbI_3$, improving the efficiency of hole-transport-layer-free solar cells [67], with an order of magnitude increase in conductivity and a factor of four increase in carrier density (from capacitance) at an incorporation of 0.03 wt% F4TCNQ.

Doping characterization and assessment

To confirm doping, photoelectron spectroscopy can be used to determine the E_F position relative to the VBM, which can be further corroborated by other measurements (for example, scanning tunnelling spectroscopy) [68].

The high-vacuum environment in photoelectron spectroscopy measurements may have an impact on the E_F position, which can shift towards the CBM, as observed for various lead HaPs [69].

VBM extraction and, thus, determination of the associated relative position to E_F, by common linear extrapolation of ultraviolet photoelectron spectroscopy data to the background level may lead to overestimation of the ionization potential, owing to low density of states near the VBM; therefore, a logarithmic detection mode and analysis have been suggested to be more appropriate for HaPs [70].

The doping efficiency in F4TCNQ-doped organic semiconductors is often assessed by the degree of charge transfer extracted from Fourier-transform infrared spectroscopy measurements [71].

Outlook

A quasi-ohmic contact created upon doping requires high local dopant densities with limited diffusion into the undoped perovskite region, which remains to be demonstrated for HaPs.

HaP p–n junctions may also be developed, assuming the possibility to strongly dope perovskites without dopant migration.

Facile ionic mobility remains a problem for perovskite doping and device design [72, 73], because it promotes defect inter-diffusion between initially p-doped and n-doped layers.

DFT calculations have shown that, for CdTe (a similar bandgap semiconductor as the most commonly explored HaPs for perovskite applications), processing under high-temperature growth conditions, which enhances defect formation, followed by rapid quenching to avoid self-compensation, allows tuning of E_F beyond the thermodynamic limit [74].

Dynamic disorder leads to the desired self-healing ability of perovskites, but complicates traditional doping, which requires kinetically stabilized defects to prevent the re-formation of the lattice.

Regulation of the Luminescence Mechanism of Two-Dimensional Tin Halide Perovskites

DOI: https://doi.org/10.1038/s41467-021-27663-0

This article is licensed under a Creative Commons Attribution 4.0 International License, which permits use, sharing, adaptation, distribution and reproduction in any medium or format, as long as you give appropriate credit to the original author(s) and the source, provide a link to the Creative Commons license, and indicate if changes were made. The images or other third party material in this article are included in the article's Creative Commons license, unless indicated otherwise in a credit line to the material. If material is not included in the article's Creative Commons license and your intended use is not permitted by statutory regulation or exceeds the permitted use, you will need to obtain permission directly from the copyright holder. To view a copy of this license, visit http://creativecommons.org/licenses/by/4.0/.

Originators: Zhang, T., Zhou, C., Feng, X. *et al.* Regulation of the luminescence mechanism of two-dimensional tin halide perovskites. *Nat Commun* 13, 60 (2022). https://doi.org/10.1038/s41467-021-27663-0

Abstract-Summary

Two-dimensional (2D) Sn-based perovskites are a kind of non-toxic environment-friendly luminescent material.

It is revealed that exciton-phonon scattering and exciton-defect scattering have different effects on exciton relaxation.

Our study further confirms that the deformation potential scattering by charged defects, not by the non-polar optical phonons, dominates the excitons interband relaxation, which is largely different from the Pb-based perovskites.

These results enhance the understanding of the origin of the non-radiative pathway in Sn-based perovskite materials.

Introduction

For the luminescence properties of 2D Pb-based perovskites, through time-resolved and temperature-dependent PL studies, Huang and others revealed that the scattering of excitons with acoustic phonons and nonpolar optical phonons is the main factor affecting the luminous efficiency of mechanically stripped 2D Pb-based perovskite sheets [75].

We note that the structural composition of the organic cationic layer can influence the defect state density and the exciton–phonon interaction [76], so we can vary the types of organic cationic layers and compare the effects of different cations on the luminescence properties and carrier dynamics of 2D Sn-based perovskites to distinguish how defects and phonon scattering affect the optical properties of materials, and furtherly reveal the difference in the factors affecting the luminescence properties between Sn-based and Pb-based perovskites.

We find it is the deformation potential scattering by charged defects, not by the optical phonons, that dominates the interband exciton relaxation, leading to an increase in the proportion of non-radiative relaxation of photogenerated carriers, for the 2D Sn-based perovskite materials.

Results and discussion

The relaxation rate of component II in PEAI-rich with smaller defect state density was slower than that of $(PEA)_2SnI_4$, which further confirmed that the effect of deformation potential scattering by charged defects on exciton interband recombination in 2D Sn-based perovskites is weakened by reducing the density of defect states.

They can reduce the fluctuation of the inorganic layer structure to weaken the scattering effect between excitons and optical phonons, thus reducing the FWHM of the PL and stably maintaining its luminous color in the temperature range of 77–300 K with no structural phase transition lower than 200 K. Second, through the TA experiment, we discovered that it is the deformation potential scattering by charged defects, not by the optical phonons, that dominates the interband excitons relaxation, diminishing the PLQY for the 2D Sn-based perovskite materials.

Methods

XPS was performed on the polycrystalline thin films prepared on Si substrates using a Thermo Scientific K-Alpha photoelectron spectrometer with Al Kα radiation (hv = 1486.6 eV).

UPS was performed on the polycrystalline thin films prepared on Si substrates by Thermo Scientific Escalab 250Xi system with an hv = 21.22 eV under an applied negative bias of 5.0 V and pass energy of 2.0 eV. For temperature-dependent PL measurement, polycrystalline thin films prepared on quartz substrates were mounted in a cryostat (Janis ST-100) and cooled by liquid nitrogen.

PLQY of polycrystalline thin films prepared on quartz substrates was measured using the Edinburgh FLS1000 instrument with an excitation wavelength of 520 nm at 300 K. Raman spectra of polycrystalline thin films prepared on quartz substrates were obtained with a Raman spectrometer (LHA19120048) using a CW laser (325 nm) as the emission source at room temperature.

Structural, Electronic, and Optical Properties of Lower-Dimensional Hybrid Perovskite Lead-Iodide Frameworks + SOC via Density Functional Theory

DOI: https://doi.org/10.1007/s42247-023-00484-1

Originators: Yami, N.F.N.A., Ramli, A., Nawawi, W.I. *et al.* Structural, electronic, and optical properties of lower-dimensional hybrid perovskite lead-iodide frameworks + SOC via density functional theory. *emergent mater.* 6, 999–1007 (2023). https://doi.org/10.1007/s42247-023-00484-1

Abstract-Summary

To create a 2D structure with high-efficiency properties, the A cation in the parental 3D structure should be replaced with a bulky organic cation (BOC).

In this study, we aim to investigate the structural, electrical, and optical characteristics of 2D (2-AMP)PbI_4 via CASTEP computer code and density functional theory (DFT).

Characteristics of GGA-PBEsol demonstrate great agreement with experiment data.

When it comes to optical characteristics, the actual part of the dielectric function reveals that this compound exhibits plasmonic behavior, which increases its capacity to absorb light.

The absorption coefficient of (2-AMP)PbI_4 shows that this 2D compound able to absorb light in the range of UV and visible light, making it a possible candidate for high-efficiency solar cell devices.

Introduction

The general formula for 3-dimensional (3D) metal halide perovskite can be expressed as ABX_3 (where A = methylammonium (MA) or formamidinium (FA), B = metal cation such as Pb^{2+} and Sn^{2+}, X = halide anion such as Br^-, Cl^-, I^-) [77].

Metal halide perovskites have recently emerged as a highly promising group of functional materials with a wide range of uses such as optoelectronic devices, including light-emitting diodes, transistors, and lasing applications, as well as other fascinating electronic features [78].

AMP, also known as (aminomethyl)pyridine, with the chemical formula of $C_6H_8N_2^+$, is one type of BOC used to fabricate of some hybrid metal halide perovskite, which can be utilized in solar cells.

To 3D counterparts, 2D perovskite structure with the general formula $(R\text{-}NH_3)_2BX_4$ or $(H_3N\text{-}R\text{-}NH_3)BX_4$ (R = alkyl chain, B = divalent metal, X = halide) has great environmental stability since the bulky hydrophobic organic cationic chains between inorganic octahedral cages may prevent moisture and oxygen from infiltrating the materials [79].

Computational details

The Cambridge Series Total Energy Package (CASTEP) code in Materials Studio 8.0 software is used for calculation, based on density functional theory (DFT).

All calculations used the ultrasoft pseudopotential and zeroth-order regular approximation (ZORA) as a relativistic treatment in heavy elements such as lead and iodine.

The plane-wave cut-off energy of 380 eV and Monkhorst–Pack k-points $1 \times 2 \times 1$ was employed.

The relativistic effect of lead was considered to include SOC as a correction to band gap calculation.

Results and discussions

This value indicates that GGA-PBEsol exhibits the best agreement with crystallographic parameters, with the percentage difference being 3.18% which is close to experimental work compared to others functional.

To better understand the electronic properties of our investigated compound (2-AMP)PbI_4, the electronic band structure was calculated by using GGA-PBEsol functional with and without SOC.

This demonstrates that Pb has a strong SOC effect which reduces CBM energy and leads result to be lower than the experimental value.

The relationship between the optical properties for dielectric function resulting from the impact of light traveling through the materials is as follows:where ε_1 (ω) and ε_2 (ω) are the real and imaginary parts, respectively.

Based on the optical result, the 2D compound of (2-AMP)PbI_4 can absorb light in UV and visible region, making it a possible candidate for high-efficiency solar cell devices.

Conclusion

DFT was used to investigate the structural, electrical, and optical aspects of (2-AMP)PbI_4.

Regarding structural properties, the lattice parameter of (2-AMP)PbI_4 calculated with GGA-PBEsol was in excellent agreement with experimental data, with a percentage error of less than 4%.

Analysis of electronic properties can be concluded that the band gap of (2-AMP)PbI_4 calculated using GGA-PBEsol, which is 1.92 eV, is close to experimental work (1.96 eV), while for band gap calculated using GGA-PBEsol + SOC demonstrated a decrease electronic structure in CBM due to splitting of Pb-p states.

For optical properties, the real part of the dielectric function shows this compound has plasmonic behavior which increases the ability of our compound to absorb light.

Bibliography

1. Mitzi, D.B., Feild, C.A., Harrison, W.T.A., Guloy, A.M.: Conducting tin halides with a layered organic-based perovskite structure. Nature **369**, 467–469 (1994)
2. Ishihara, T., Takahashi, J., Goto, T.: Optical properties due to electronic transitions in two-dimensional semiconductors (C_nH_{2n}+1NH_3)Pb_{I4}. Phys. Rev. B **42**, 11099–11107 (1990)
3. Tanaka, K., et al.: Electronic and excitonic structures of inorganic-organic perovskite-type quantum-well crystal $(C_4H_9NH_3)_2PbBr_4$. Jpn. J. Appl. Phys.**44**, 5923–5932 (2005)
4. Saparov, B., Mitzi, D.B.: Organic–inorganic perovskites: structural versatility for functional materials design. Chem. Rev. **116**, 4558–4596 (2016)
5. Shi, E., et al.: Two-dimensional halide perovskite nanomaterials and heterostructures. Chem. Soc. Rev. **47**, 6046–6072 (2018)

6. Tsai, H., et al.: High-efficiency two-dimensional Ruddlesden-Popper perovskite solar cells. Nature **536**, 312–316 (2016)
7. Blancon, J.-C., et al.: Extremely efficient internal exciton dissociation through edge states in layered 2D perovskites. Science **355**, 1288–1292 (2017)
8. Dou, L., et al.: Atomically thin two-dimensional organic-inorganic hybrid perovskites. Science **349**, 1518–1521 (2015)
9. Niu, W., Eiden, A., Prakash, G.V., Baumberg, J.J.: Exfoliation of self-assembled 2D organic–inorganic perovskite semiconductors. Appl. Phys. Lett. **104**, 171111 (2014)
10. Mitzi, D.B.: In: Karlin, K. (eds.) Progress in Inorganic Chemistry, vol. 48, pp. 1–121. Wiley, New York, NY (1999)
11. Kresse, G., Hafner, J.: Ab initio molecular dynamics for liquid metals. Phys. Rev. B **47**, 558–561 (1993)
12. Furchi, M.M., Polyushkin, D.K., Pospischil, A., Mueller, T.: Mechanisms of photoconductivity in atomically thin MoS_2. Nano Lett. **135**, 2089–2096 (2015)
13. Tan, Z., Wu, Y., Hong, H., Yin, J., Zhang, J., Lin, L., Wang, M., Sun, X., Sun, L., Huang, Y., Liu, K., Liu, Z., Peng, H.: Two-dimensional $(C_4H_9NH_3)_2PbBr_4$ perovskite crystals for high-performance photodetector. J. Am. Chem. Soc. **138**(51), 16612–16615 (2016). https://doi.org/10.1021/jacs.6b11683
14. Blancon, J.-C., et al.: The effects of electronic impurities and electron–hole recombination dynamics on large-grain organic–inorganic perovskite photovoltaic efficiencies. Adv. Funct. Mater.Funct. Mater. **26**, 4283–4292 (2016)
15. Stranks, S.D., et al.: Electron–hole diffusion lengths exceeding 1 micrometer in an organometal trihalide perovskite absorber. Science **342**, 341–344 (2013)
16. Miyata, A., et al.: Direct measurement of the exciton binding energy and effective masses for charge carriers in organic-inorganic trihalide perovskites. Nat. Phys. **11**, 582–587 (2015)
17. Nie, W., et al.: Critical role of interface and crystallinity on the performance and photostability of perovskite solar cell on nickel oxide. Adv. Mater. **30**, 1703879 (2018)
18. Nie, W., et al.: Light-activated photocurrent degradation and self-healing in perovskite solar cells. Nat. Commun. **7**, 11574 (2016)
19. Tsai, H., et al.: Optimizing composition and morphology for large-grain perovskite solar cells via chemical control. Chem. Mater. **27**, 5570–5576 (2015)
20. Nie, W., et al.: High-efficiency solution-processed perovskite solar cells with millimeter-scale grains. Science **347**, 522–525 (2015)
21. Tsai, H., et al.: Light-induced lattice expansion leads to high-efficiency and stable perovskite solar cells. Science **360**, 67–70 (2018)
22. Hanamura, E., Nagaosa, N., Kumagai, M., Takagahara, T.: Quantum wells with enhanced exciton effects and optical non-linearity. Mater. Sci. Eng. B **1**, 255–258 (1988)
23. Hong, X., Ishihara, T., Nurmikko, A.: Dielectric confinement effect on excitons in PbI_4-based layered semiconductors. Phys. Rev. B **45**(12), 6961 (1992). https://doi.org/10.1103/PhysRevB.45.6961
24. Ishihara, T., Takahashi, J., Goto, T.: Exciton state in two-dimensional perovskite semiconductor $(C_{10}H_{21}NH_3)_2PbI_4$. Solid State Commun. **69**, 933–936 (1989)
25. Pedesseau, L., et al.: Advances and promises of layered halide hybrid perovskite semiconductors. ACS Nano **10**, 9776–9786 (2016)
26. Milot, R.L., et al.: Charge-carrier dynamics in 2D hybrid metal-halide perovskites. Nano Lett. **16**, 7001–7007 (2016)
27. Mercier, N., et al.: Unique hydrogen bonding correlating with a reduced band gap and phase transition in the hybrid perovskites $(HO(CH_2)2NH_3)_2PbX_4$ (X = I, Br). Inorg. Chem.**43**, 8361–8366 (2004)
28. Covington, A.K., Dickinson, T.: Physical Chemistry of Organic Solvent Systems. Plenum, New York (1973)
29. Ren, K., et al.: Turning a disadvantage into an advantage: synthesizing high-quality organometallic halide perovskite nanosheet arrays for humidity sensors. J. Mater. Chem. C. **5**, 2504–2508 (2017)

30. Quan, L.N., et al.: Perovskites for next-generation optical sources. Chem. Rev. **119**, 7444–7477 (2019)
31. Wu, X., Trinh, M.T., Zhu, X.Y.: Excitonic many-body interactions in two-dimensional lead iodide perovskite quantum wells. J. Phys. Chem. C **119**, 14714–14721 (2015)
32. Hill, A.H., Kennedy, C.L., Massaro, E.S., Grumstrup, E.M.: Perovskite carrier transport: disentangling the impacts of effective mass and scattering time through microscopic optical detection. J. Phys. Chem. Lett. **9**, 2808–2813 (2018)
33. Pazos-Outón, L.M., et al.: Photon recycling in lead iodide perovskite solar cells. Science **351**, 1430–1433 (2016)
34. Gan, Z., et al.: The dominant energy transport pathway in halide perovskites: photon recycling or carrier diffusion? Adv. Energy Mater. **9**, 1900185 (2019)
35. Bercegol, A., et al.: Quantitative optical assessment of photonic and electronic properties in halide perovskite. Nat. Commun. **10**, 1–8 (2019)
36. Wang, F., Wu, Y., Hybertsen, M.S., Heinz, T.F.: Auger recombination of excitons in one-dimensional systems. Phys. Rev. B **73**, 245424 (2006)
37. Philbin, J.P., Rabani, E.: Electron-hole correlations govern Auger recombination in nanostructures. Nano Lett. **18**, 7889–7895 (2018)
38. Ma, X., et al.: Influences of exciton diffusion and exciton-exciton annihilation on photon emission statistics of carbon nanotubes. Phys. Rev. Lett. **115**, 017401 (2015)
39. Hou, X., et al.: Engineering Auger recombination in colloidal quantum dots via dielectric screening. Nat. Commun. **10**, 1750 (2019)
40. Mitzi, D.B., Chondroudis, K., Kagan, C.R.: Design, structure, and optical properties of organic-inorganic perovskites containing an oligothiophene chromophore. Inorg. Chem. **38**, 6246–6256 (1999)
41. Passarelli, J.V., et al.: Enhanced out-of-plane conductivity and photovoltaic performance in n = 1 layered perovskites through organic cation design. J. Am. Chem. Soc. **140**, 7313–7323 (2018)
42. Van Gompel, W.T.M., et al.: Towards 2D layered hybrid perovskites with enhanced functionality: introducing charge-transfer complexes via self-assembly. Chem. Commun. **55**, 2481–2484 (2019)
43. Mitzi, D.B., Medeiros, D.R., Malenfant, P.R.L.: Intercalated organic-inorganic perovskites stabilized by fluoroaryl-aryl interactions. Inorg. Chem.**41**, 2134–2145 (2002)
44. Gao, X., et al.: Ruddlesden-Popper perovskites: synthesis and optical properties for optoelectronic applications. Adv. Sci. **6**, 1900941 (2019)
45. Mao, L., et al.: Hybrid Dion-Jacobson 2D lead iodide perovskites. J. Am. Chem. Soc. **140**, 3775–3783 (2018)
46. Ahmad, S., et al.: Dion-Jacobson phase 2D layered perovskites for solar cells with ultrahigh stability. Joule **3**, 794–806 (2019)
47. Lin, J., et al.: Thermochromic halide perovskite solar cells. Nat. Mater. **17**, 261–267 (2018)
48. Wheeler, L.M., et al.: Switchable photovoltaic windows enabled by reversible photothermal complex dissociation from methylammonium lead iodide. Nat. Commun. **8**, 1722 (2017)
49. Burger, S., Ehrenreich, M.G., Kieslich, G.: Tolerance factors of hybrid organic–inorganic perovskites: recent improvements and current state of research. J. Mater. Chem. A **6**, 21785–21793 (2018)
50. Bartel, C.J., et al.: New tolerance factor to predict the stability of perovskite oxides and halides. Sci. Adv. **5**, eaav0693 (2019)
51. Li, Z., et al.: Stabilizing perovskite structures by tuning tolerance factor: formation of formamidinium and cesium lead iodide solid-state alloys. Chem. Mater. **28**, 284–292 (2015)
52. Chuntonov, L., Kumar, R., Kuroda, D.G.: Non-linear infrared spectroscopy of the water bending mode: direct experimental evidence of hydration shell reorganization? Phys. Chem. Chem. Phys. **16**, 13172–13181 (2014)
53. Coker, D.F., Reimers, J.R., Watts, R.O.: The infrared absorption spectrum of water. Aust. J. Phys. **35**, 623–638 (1982)

54. Gélvez-Rueda, M.C., et al.: Interconversion between free charges and bound excitons in 2D hybrid lead halide perovskites. J. Phys. Chem. C **121**, 26566–26574 (2017)
55. Zhang, X., Li, L., Sun, Z., Luo, J.: Rational chemical doping of metal halide perovskites. Chem. Soc. Rev. **48**, 517–539 (2019)
56. Zhou, Y., Chen, J., Bakr, O.M., Sun, H.-T.: Metal-doped lead halide perovskites: synthesis, properties, and optoelectronic applications. Chem. Mater. **30**, 6589–6613 (2018)
57. Mitzi, D.B., Feild, C.A., Schlesinger, Z., Laibowitz, R.B.: Transport, optical, and magnetic properties of the conducting halide perovskite $CH_3NH_3SnI_3$. J. Solid State Chem. **114**, 159–163 (1995)
58. Takahashi, Y., et al.: Tunable charge transport in soluble organic–inorganic hybrid semiconductors. Chem. Mater. **19**, 6312–6316 (2007)
59. Takahashi, Y., et al.: Charge-transport in tin-iodide perovskite CH3NH3SnI3: origin of high conductivity. Dalton Trans. **40**, 5563–5568 (2011)
60. Takahashi, Y., Hasegawa, H., Takahashi, Y., Inabe, T.: Hall mobility in tin iodide perovskite $CH_3NH_3SnI_3$: evidence for a doped semiconductor. J. Solid State Chem. **205**, 39–43 (2013)
61. Haque, M.A.: Tuning the thermoelectric performance of hybrid tin perovskites by air treatment. Adv. Energy Sustain. Res. **1**, 2000033 (2020)
62. Sze, S.M., Ng, K.K.: Physics of Semiconductor Devices. Wiley (2006).
63. Streetman, B.G., Banerjee, S.: Solid State Electronic Devices, 5th edn (2000)
64. Xu, P., Chen, S., Xiang, H.J., Gong, X.G., Wei, S.H.: Influence of defects and synthesis conditions on the photovoltaic performance of perovskite semiconductor $CsSnI_3$. Chem. Mater. **26**, 6068–6072 (2014)
65. Shi, T., et al.: Effects of organic cations on the defect physics of tin halide perovskites. J. Mater. Chem. A **5**, 15124–15129 (2017)
66. Meggiolaro, D., Ricciarelli, D., Alasmari, A.A., Alasmary, F.A.S., De Angelis, F.: Tin versus lead redox chemistry modulates charge trapping and self doping in tin/lead-iodide perovskites. J. Phys. Chem. Lett. **11**, 3546–3556 (2020)
67. Wu, W., et al.: Molecular doping enabled scalable blading of efficient hole-transport-layer-free perovskite solar cells. Nat. Commun. **9**, 1625 (2018)
68. Kahn, A.: Fermi level, work function and vacuum level. Mater. Horiz.**3**, 7–10 (2016)
69. Hu, Z., et al.: The impact of atmosphere on energetics of lead halide perovskites. Adv. Energy Mater. **10**, 2000908 (2020)
70. Zhang, F., et al.: Ultraviolet photoemission spectroscopy and Kelvin probe measurements on metal halide perovskites: advantages and pitfalls. Adv. Energy Mater. **10**, 1903252 (2020)
71. Méndez, H., et al.: Charge-transfer crystallites as molecular electrical dopants. Nat. Commun.**6**, 8560 (2015)
72. Eames, C., et al.: Ionic transport in hybrid lead iodide perovskite solar cells. Nat. Commun.**6**, 7497 (2015)
73. Game, O.S., Buchsbaum, G.J., Zhou, Y., Padture, N.P., Kingon, A.I.: Ions matter: Description of the anomalous electronic behavior in methylammonium lead halide perovskite devices. Adv. Funct. Mater. **27**, 1606584 (2017)
74. Yang, J.H., et al.: Tuning the fermi level beyond the equilibrium doping limit through quenching: the case of CdTe. Phys. Rev. B Condens. Matter Mater. Phys. **90**, 245202 (2014)
75. Guo, Z., Wu, X., Zhu, T., Zhu, X., Huang, L.: Electron–phonon scattering in atomically thin 2D perovskites. ACS Nano **10**, 9992–9998 (2016)
76. Liao, Y., et al.: Highly oriented low-dimensional tin halide perovskites with enhanced stability and photovoltaic performance. J. Am. Chem. Soc. **139**, 6693–6699 (2017)
77. Seo, J., Noh, J.H., Seok, S.I: Rational strategies for efficient perovskite solar cells. Acc. Chem. Res. **49**(3), 562–572 (2016). https://doi.org/10.1021/acs.accounts.5b00444. (American Chemical Society)
78. Dong, H., Ran, C., Gao, W., Li, M., Xia, Y., Huang, W.: Metal halide perovskite for next-generation optoelectronics: progresses and prospects. eLight **3**(1), 3 (2023). https://doi.org/10.1186/s43593-022-00033-z

79. Xu, T., Xiang, W., Kubicki, D.J., Liu, Y., Tress, W., Liu, S.: Simultaneous lattice engineering and defect control via cadmium incorporation for high-performance inorganic perovskite solar cells. Adv. Sci. **9**(36), 2204486 (2022). https://doi.org/10.1002/advs.202204486

Chapter 3
Lattice Properties

Introduction by Editor

The metal halide lattice plays a large role in dictating the optoelectronic properties of MHPs. The lattice is very ionic in nature and therefore couples strongly to the charge carriers. The nature of the charge carrier, whether an exciton or a free electron/hole, also dictates the type of carrier-phonon interaction. In 3D MHPs with low exciton binding energies, the carriers exist as large polarons which can move freely but are dressed with phonon clouds and interact with the lattice through the Fröhlich Hamiltonian. On the other hand, in 2D MHPs with large exciton binding energies, carriers take the form of small polarons and couple to the lattice through Holstein-like interactions.

The articles in this chapter are presented in chronological order. Two of the articles, "Electron–phonon coupling in hybrid lead halide perovskites" and "Electron-phonon interactions in halide perovskites", primarily discuss the lattice properties of 3D MHPs. We felt it prudent to include these articles since the physical principles discussed in these articles also apply to 2D MHPs—the phonon structure of 2D and 3D MHPs are similar, comprised mainly of ~ 1 to 5 meV LO phonons in the metal-halide framework and with an octahedral-tilting mode that couples strongly to charge carriers.

It's worth noting that defect densities in 2D and 3D MHPs are orders of magnitude higher than in traditional semiconductors such as Si or GaAs. If such high defect densities were realized within a Si- or GaAs-based device, it would ruin the device performance. Thus, the high performance of MHPs notwithstanding their defect-riddled lattices is remarkable and MHPs are among the most defect-tolerant semiconductors. Some have hypothesized that the polaronic nature of charge-carriers is a critical mechanism underlying the defect tolerance of MHPs, arguing that the phonon cloud dressing the carrier efficiently screens its interactions with defects and

© The Author(s), under exclusive license to Springer Nature Singapore Pte Ltd. 2024
J. S. Colton and K. R. Hansen (eds.), *Two-Dimensional Metal Halide Perovskites*,
https://doi.org/10.1007/978-981-99-7830-4_3

extends the carrier lifetimes. Meanwhile, others argue that a dynamic Rashba effect is responsible for the long carrier lifetimes. As argued, the dynamic motion of the lattice causes local symmetry breaking. This induces a Rashba effect, splitting the conduction band into two bands offset from the valance band maximum in k-space, thereby making radiative recombination a momentum-limited transition.

The interatomic potentials in MHPs are highly anharmonic which is evidenced by the fact that many MHP crystals undergo multiple phase transitions. This deviation from harmonic behavior complicates the study of polaronic effects since the established polaron theories are based on the harmonic approximation. Thus, in many respects MHPs are ideal templates for investigating complex charge-lattice interactions, since these materials challenge traditional assumptions.

Both 3D and 2D MHP lattices are dynamically active at room temperature. Notably, halide atoms have been observed to migrate through lattice vacancies, which phenomenon is known as ion migration. Meanwhile, the A-site molecules are free to rotate as they only interact with the lattice through relatively weak hydrogen bonds. The motion of A-site molecules imparts dynamic disorder and local potential fluctuations on band-edge states and molecular dynamics simulations have demonstrated this effect to be substantial.

Charge-lattice interactions in 2D MHPs are even evident in simple optical absorption experiments. The excitonic absorption below the bandgap exhibits multiple peaks that are a result of charge-lattice interactions; however, the precise nature of the charge-lattice interaction remains an open question. Originally, the multiplicity of peaks was attributed to Franck–Condon progression resulting from coupling with a ~40 meV vibrational mode on a phenethylammonium cation (in e.g. phenethylammonium lead iodide). However, as highlighted in this chapter, Thouin et al. argue the origin of the peaks lies in a coexistence of multiple exciton-polaron states, each with its unique coupling to lower-energy (~5 meV) phonon modes within the metal halide framework. Still, this remains a question of considerable debate in the literature and is a very intriguing fundamental question.

Machine generated summaries

Disclaimer: The auto-summaries have been generated by a recursive clustering algorithm via the Dimensions Auto-summarizer by Digital Science handled by Subject Matter Experts and the external editor. The editor of this book selected which SN content should be auto-summarized and decided its order of appearance. Please be aware that the auto-summaries consist of original sentences, but are not representative of its original paper, since we do not show the full length of the publication. Please note that only published SN content is represented here, and that machine-generated books are still at an experimental stage.

Machine generated keywords: electronphonon, phonon, coupling, polaron, dynamic, spacer, electronphonon interaction, exciton, lattice, chargecarrier, interaction, character, interlayer, excitonic, distance

Electron–Phonon Coupling in Hybrid Lead Halide Perovskites

DOI: https://doi.org/10.1038/ncomms11755

This work is licensed under a Creative Commons Attribution 4.0 International License. The images or other third party material in this article are included in the article's Creative Commons license, unless indicated otherwise in the credit line; if the material is not included under the Creative Commons license, users will need to obtain permission from the license holder to reproduce the material. To view a copy of this license, visit http://creativecommons.org/licenses/by/4.0/

Originators: Wright, A., Verdi, C., Milot, R. *et al.* Electron–phonon coupling in hybrid lead halide perovskites. *Nat Commun* 7, 11755 (2016). https://doi.org/10.1038/ncomms11755

Abstract-Summary

Phonon scattering limits charge-carrier mobilities and governs emission line broadening in hybrid metal halide perovskites.

Establishing how charge carriers interact with phonons in these materials is therefore essential for the development of high-efficiency perovskite photovoltaics and low-cost lasers.

We investigate the temperature dependence of emission line broadening in the four commonly studied formamidinium and methylammonium perovskites, $HC(NH_2)_2PbI_3$, $HC(NH_2)_2PbBr_3$, $CH_3NH_3PbI_3$ and $CH_3NH_3PbBr_3$, and discover that scattering from longitudinal optical phonons via the Fröhlich interaction is the dominant source of electron–phonon coupling near room temperature, with scattering off acoustic phonons negligible.

Introduction

These findings have therefore raised the puzzling question of why such hybrid perovskites appear to evade the Fröhlich interactions between charge carriers and polar longitudinal optical (LO) phonon modes that normally govern polar inorganic semiconductors, such as GaAs [1, 1], at room temperature.

We clarify the relative activity of different charge-carrier scattering mechanisms in hybrid lead halide perovskites by investigating charge-carrier scattering through an analysis of the photoluminescence (PL) linewidth as a function of temperature between 10 and 370 K. By carefully examining the low-temperature regime in which thermal energies fall below those of high-energy LO phonon modes, we are able to clearly separate competing contributions from charge-carrier interactions with acoustic and optical phonons.

Our results conclusively demonstrate that electron–phonon coupling in hybrid lead halide perovskites follows a classic bandstructure picture for polar inorganic semiconductors, which are dominated by Fröhlich coupling between charge carriers and LO phonon modes in the high-temperature regime.

Results

The lack of uncomplicated low-temperature spectra has to date prevented proper analysis of the linewidth broadening of MA perovskites and the associated phonon coupling to charge carriers.

Both methods yield upper limits around γ_{ac}=60±20 μeV K^{-1} for $FAPbI_3$ and $FAPbBr_3$; therefore, acoustic phonons will only contribute up to ~18 meV to the linewidth broadening at 300 K. Hence, our analysis illustrates that the majority of broadening in the room-temperature regime arises from charge-carrier interactions with optical phonons, as would be expected for a polar semiconductor.

We have recently examined infrared photoinduced transmission spectra for methylammonium lead iodide perovskite [2] and found these to be predominantly governed by free-charge (Drude-like) features, with localization effects (for example, from excitons) only contributing at low temperature and not more than around 23% even at 8 K. This suggests that over the temperature window we examine here, emission broadening is mostly governed by interactions between phonons and free charge carriers rather than excitons.

Discussion

Our analysis of the experimentally determined emission linewidth broadening and our first-principles calculations strongly support the notion that Fröhlich coupling to LO phonons is the predominant charge-carrier scattering mechanism in hybrid lead halide perovskites.

We conclude that Fröhlich coupling to LO phonons, rather than acoustic phonon deformation potential coupling, is the dominant charge-carrier scattering mechanism at room temperature in these hybrid lead halide perovskites.

Our analysis of the temperature-dependent emission linewidth of $FAPbI_3$ and $FAPbBr_3$ allowed us to establish that the Fröhlich interaction between charge carriers and LO phonons provides the dominant contribution to the predominantly homogeneous linewidth broadening in these hybrid perovskites at room temperature.

We furthermore obtained experimentally measured energies of LO phonon modes responsible for Fröhlich interactions in these materials and showed that Fröhlich interactions are higher for bromide perovskites than iodide perovskites, providing a link between composition and electron–phonon scattering that fundamentally limits charge-carrier motion.

Methods

MA perovskite films were deposited in a nitrogen-filled glovebox using a solvent quenching method wherein an excess of antisolvent is deposited onto the wet substrate while spin-coating [3].

A 1:1 molar ratio solution of MAX and PbX_2 (X=I, Br) was dissolved in anhydrous N,N-dimethylformamide at 1 M. This was then spin-coated onto the quartz substrates at 5,000 r.p.m.

FAX and PbX_2 (X=I, Br) were dissolved in anhydrous N,N-dimethylformamide in a 1:1 molar ratio at 0.55 M. Immediately before film formation, small amounts of acid were added to the precursor solutions, to enhance the solubility of the precursors and allow smooth and uniform film formation.

Films were then spin coated from the precursor plus acid solution on warm (85 °C) oxygen plasma-cleaned substrates at 2,000 r.p.m.

Electron–Phonon Interaction in Efficient Perovskite Blue Emitters

DOI: https://doi.org/10.1038/s41563-018-0081-x

Originators: Gong, X., Voznyy, O., Jain, A. *et al.* Electron–phonon interaction in efficient perovskite blue emitters. *Nature Mater* 17, 550–556 (2018). DOI: https://doi.org/10.1038/s41563-018-0081-x

Abstract-Summary

We investigate the effect of electron–phonon interactions on the luminescence of single crystals of two-dimensional perovskites, showing that reducing these interactions can lead to bright blue emission in two-dimensional perovskites.

By varying the molecular configuration of the ligands, we show that a PLQY up to 79% and linewidth of 20 nm can be reached by controlling crystal rigidity and electron–phonon interactions.

Designing crystal structures with electron–phonon interactions in mind offers a previously underexplored avenue to improve optoelectronic materials' performance.

Main

We sought therefore to explore further the potential of 2D perovskite blue emitters using high-quality single crystals, and investigate the determinants of high brightness and thus offer an improved set of guidelines in materials design.

Building these macroscopic, atomically layered sheet structures required us to develop a new synthesis strategy as high-brightness macroscale 2D perovskite single crystals have not previously been reported.

We carried out single-crystal X-ray diffraction (SCXRD) on samples with high PLQY (PhC2), low PLQY (C4) and broadband emission spectra (Ph).

To gain further insight into the crystal quality and high PLQY, we carried out density functional theory (DFT) calculations on C4, Ph and PhC2 structures.

This work reports a deep-blue emitter with a high PLQY based on high-quality 2D perovskite single crystals.

Methods

Samples for all PL measurements were prepared by mechanical exfoliation from single crystals.

The incident and PL spectra were collected with the exfoliated sample (tape/exfoliated crystals/glass substrate) directly in the excitation beam path offset from the beam path and removed from the sphere, in accordance with the standard method.

3M Scotch tape absorbs 15% incident light (350 nm excitation) before it reaches the exfoliated crystals, the effect of which was calibrated by multiplying the measured PLQY by $1/(1 - 0.15) = 1.17$.

In the sample of exfoliated PhC2 single crystals, the PL intensity of PhC2 slightly increased from 10 K to 150 K, which might originate from gradual thermal activation of the bright excitonic state.

Structural distortion does not vary with temperature: the Br–Pb–Br bond angle was measured as 168–172° (PhC2) and 180° (C4) at 100 K. The DFT calculations also show that at 0 K the corresponding angle for PhC2 is 168–170° and that for C4 is 180°.

Cross-Plane Coherent Acoustic Phonons in Two-Dimensional Organic-Inorganic Hybrid Perovskites

DOI: https://doi.org/10.1038/s41467-018-04429-9

This article is licensed under a Creative Commons Attribution 4.0 International License, which permits use, sharing, adaptation, distribution and reproduction in any medium or format, as long as you give appropriate credit to the original author(s) and the source, provide a link to the Creative Commons license, and indicate if changes were made. The images or other third party material in this article are included in the article's Creative Commons license, unless indicated otherwise in a credit line to the material. If material is not included in the article's Creative Commons license and your intended use is not permitted by statutory regulation or exceeds the permitted use, you will need to obtain permission directly from the copyright holder. To view a copy of this license, visit http://creativecommons.org/licenses/by/4.0/.

Originators: Guo, P., Stoumpos, C.C., Mao, L. *et al.* Cross-plane coherent acoustic phonons in two-dimensional organic-inorganic hybrid perovskites. *Nat Commun* 9, 2019 (2018). https://doi.org/10.1038/s41467-018-04429-9

Abstract-Summary

Two-dimensional Ruddlesden–Popper organic–inorganic hybrid layered perovskites (2D RPs) are solution-grown semiconductors with prospective applications in next-generation optoelectronics.

We report on the generation and propagation of coherent longitudinal acoustic phonons along the cross-plane direction of 2D RPs, following separate characterizations of below-bandgap refractive indices.

Through experiments on single crystals of systematically varied perovskite layer thickness, we demonstrate significant reduction in both group velocity and propagation length of acoustic phonons in 2D RPs as compared to the three-dimensional methylammonium lead iodide counterpart.

Introduction

2D RPs additionally present features of superlattices, an attractive material class that can exhibit desirable thermal properties that differ from the bulk limits [8] while still preserving the wave nature of phonons [9, 10].

Unlike those analogs, 2D RPs can be grown into single crystals with tens-to-hundreds-of-micron cross-plane dimensions, offering an archetypal system for the study of intrinsic phonon transport characteristics.

We find 2D RPs exhibit significantly lower group velocities compared to the 3D counterpart, methylammonium lead iodide ($MAPbI_3$), which corresponds to $N = \infty$. As captured by a coarse-grained bead-spring model, the reduction in phonon velocity originates from the weak van der Waals bonding between the organic spacers.

The unusually large acoustic impedance mismatch between the organic and inorganic layers, and the loose $CH_3(CH_2)_m$ aliphatic tails of the organic spacer layers with partial liquid-like motions lead to substantially stronger suppression of acoustic phonon propagation in 2D RPs as compared to $MAPbI_3$.

Results

For any given member, the small variation of k_b (<10%) relative to the variation of k_a (13–22 N m^{-1}) stems from the order-of-magnitude difference between k_b and k_a; such difference dictates that the CLAP velocity is mainly determined by k_b, while a relatively large variation of k_a does not significantly alter the velocity.

The significant difference between k_c and k_d clearly demonstrates that the slow cross-plane acoustic phonon velocities observed for $N = 1$ to $N = 6$ primarily stem from the weak van der Waals interactions between the organic spacer layers, whereas the organic–inorganic interface only plays a secondary role.

That the observed variation of oscillation amplitude with N does not mainly arise from a difference in the magnitude of index modulation due to a bandgap change (caused by CLAP-induced strain).

Discussion

Strong anharmonicity at the organic–inorganic interfaces stemming from an asymmetric potential landscape may also contribute to phonon scattering (e.g., stronger interfacial reflection and phonon–phonon interactions) [11] and suppress cross-plane phonon propagation [12, 13], which results in the overall short coherence times of the 2D RPs in comparison to 3D $MAPbI_3$.

Based on the stronger CLAP attenuation, we expect κ of 2D RPs to diminish more strongly than v itself, and heat dissipation of 2D RP-based devices presents an important design consideration.

We find reduced group velocities and substantially stronger attenuation of acoustic phonons in 2D RPs as compared to $MAPbI_3$, which arises from the large impedance mismatch between the organic and inorganic layers due to the weak van der Waals bonding between the layers.

2D perovskites without van der Waals bonding (e.g., (HA)PbI_4) are expected to exhibit higher thermal conductivities for device applications, yet still maintaining strong quantum confinement effects.

Methods

PbO powder (4464 mg, 20 mmol) was dissolved in a mixture of 57% w/w aqueous HI solution (25.0 ml, 190 mmol) and 50% aqueous H_3PO_2 (3.4 ml, 15.5 mmol), by heating to boiling under constant magnetic stirring for about 5 min, forming a bright yellow solution.

Subsequent addition of solid CH_3NH_3Cl, of 20 × (N−1)/N mmol, to the hot yellow solution initially caused the precipitation of a black powder which rapidly re-dissolved under stirring to yield a clear bright yellow solution.

In a separate beaker, n-$CH_3(CH_2)_3NH_2$, of 20-(20 × (N−1)/N) mmol, was neutralized with aqueous HI solution (57% w/w, 5 ml, 38 mmol) in an ice bath resulting in a clear, pale yellow solution.

Addition of the n-$CH_3(CH_2)_3NH_3I$ solution to the PbI_2 solution initially produced a black precipitate which was subsequently dissolved on heating the combined solution to boiling.

Phonon Coherences Reveal the Polaronic Character of Excitons in Two-Dimensional Lead Halide Perovskites

DOI: https://doi.org/10.1038/s41563-018-0262-7

Originators: Thouin, F., Valverde-Chávez, D.A., Quarti, C. *et al.* Phonon coherences reveal the polaronic character of excitons in two-dimensional lead halide perovskites. *Nat. Mater.* 18, 349–356 (2019). https://doi.org/10.1038/s41563-018-0262-7

Abstract-Summary

Hybrid organic–inorganic semiconductors feature complex lattice dynamics due to the ionic character of the crystal and the softness arising from non-covalent bonds between molecular moieties and the inorganic network.

We establish that such dynamic structural complexity in a prototypical two-dimensional lead iodide perovskite gives rise to the coexistence of diverse excitonic resonances, each with a distinct degree of polaronic character.

By means of high-resolution resonant impulsive stimulated Raman spectroscopy, we identify vibrational wavepacket dynamics that evolve along different configurational coordinates for distinct excitons and photocarriers.

Main

The distinct ionic character, and the 'softness' of the lattice give rise to strong electron–phonon coupling and dynamic disorder effects, which influence the optical and electronic properties of this class of materials [15–17], and their consequences on excitonic structure are not thoroughly explored.

To account for this structure, we have invoked a general framework of a delocalized Wannier exciton with binding energy ~200 meV but with substantially large coupling to local lattice vibrations, modelled by four distinct, non-degenerate excitonic transitions, spectrally separated by multiples of 35 meV [18].

We now return to the vibrational coherences observed in $(PEA)_2PbI_4$ to demonstrate contrasting coupling of the lattice to the two most intense excitonic transitions A and B, and also to photocarriers.

While the resonant excitation of exciton A reveals predominant coupling to mode M2, excitation of exciton B displays a strikingly different vibrational spectral structure.

Methods

Differential transmission spectroscopy measurements were performed using an ultrafast laser system (Pharos Model PH1-20-0200-02-10, Light Conversion) emitting 1,030-nm pulses at 100 kHz, with an output power of 20 W and pulse duration of ~220 fs.

Experiments were carried out in an integrated transient absorption/time-resolved photoluminescence commercial set-up (Light Conversion Hera).

When higher-energy probe light was required, a blue white-light continuum was similarly obtained by using the second harmonic of the laser output instead.

We disclose that we observed a slow degradation of the sample over long exposure to the laser light.

Such a process, also widely reported for 3D perovskites, appears to be reversible and can be slightly negated by photo-exposing the sample for an hour before the experiment.

The Hessian matrix of the forces is then calculated on the fully relaxed structure and diagonalized to obtain the vibrational frequencies.

Exciton Diffusion in Two-Dimensional Metal-Halide Perovskites

DOI: https://doi.org/10.1038/s41467-020-15882-w

This article is licensed under a Creative Commons Attribution 4.0 International License, which permits use, sharing, adaptation, distribution and reproduction in any medium or format, as long as you give appropriate credit to the original author(s) and the source, provide a link to the Creative Commons license, and indicate if changes were made. The images or other third party material in this article are included in the article's Creative Commons license, unless indicated otherwise in a credit line to the material. If material is not included in the article's Creative Commons license and your intended use is not permitted by statutory regulation or exceeds the permitted use, you will need to obtain permission directly from the copyright holder. To view a copy of this license, visit http://creativecommons.org/licenses/by/4.0/.

Originators: Seitz, M., Magdaleno, A.J., Alcázar-Cano, N. *et al.* Exciton diffusion in two-dimensional metal-halide perovskites. *Nat Commun* 11, 2035 (2020). https://doi.org/10.1038/s41467-020-15882-w

Abstract-Summary

The impact of the reduced dimensionality on the optoelectronic properties remains unclear, particularly regarding the spatial dynamics of the excitonic excited state within the two-dimensional plane.

We present direct measurements of exciton transport in single-crystalline layered perovskites.

Using transient photoluminescence microscopy, we show that excitons undergo an initial fast diffusion through the crystalline plane, followed by a slower subdiffusive regime as excitons get trapped.

A clear correlation between lattice stiffness and diffusivity is found, suggesting exciton–phonon interactions to be dominant in the spatial dynamics of the excitons in perovskites, consistent with the formation of exciton–polarons.

Introduction

The improved environmental stability in 2D perovskite phases is attributed to a better moisture resistance due to the hydrophobic organic spacers that passivate the inorganic perovskite sheets, as well as an increased formation energy of the material [19–22].

2D perovskites are composed of inorganic metal-halide layers, which are separated by long organic spacer molecules.

Light harvesting using 2D perovskites relies on the efficient transport of excitons and their subsequent separation into free charges [23].

While charge transport in bulk perovskites has been studied in great detail, the mechanisms that dictate exciton transport in 2D perovskites remain elusive [24].

We report the direct visualization of exciton diffusion in 2D single-crystalline perovskites using transient photoluminescence-microscopy [25].

Using the versatility of perovskite materials, we study the influence of the organic spacer on the diffusion dynamics of excitons in 2D perovskites.

Results

This strongly suggests that the early diffusion dynamics is unaffected by energetic disorder, which would result in a sublinear behavior with $\alpha < 1$, and any trap states, giving us direct access to the intrinsic exciton diffusivity of the material and allowing us to compare the intrinsic exciton diffusivity between perovskites of different compositions.

To investigate the possible role of exciton–phonon coupling on exciton diffusion, we first quantify the softness of the lattices of both $(PEA)_2PbI_4$ and $(BA)_2PbI_4$ by extracting the atomic displacement parameters from their respective single crystal X-ray data [26].

Across the entire range of organic spacers, a clear correlation between the diffusivity and the lattice softness is found, further confirming the dominant role of exciton–phonon coupling in the spatial dynamics of the excited state in 2D perovskites.

The diffusion of large polarons decreases with increasing temperature ($\partial D/\partial T < 0$), resembling that of band-like free exciton motion, although with a strongly increased effective mass.

Discussion

The observed correlation between diffusivity and lattice softness in combination with band-like transport is in good qualitative agreement with the formation of large exciton–polarons.

Taking into account the close correlation between diffusivity and the atomic displacement, this parameter space can be readily explored using available X-ray crystal structure data for many 2D perovskite analogues.

While the influence of the organic spacer is expected to be particularly strong in the class of $n = 1$ 2D perovskites, we have observed consistent trends in the $n = 2$ analogues.

We have studied the spatial and temporal exciton dynamics in 2D metal-halide perovskites of the form L_2PbI_4.

Our results indicate that exciton–phonon interactions dominate the spatial dynamics of excitons in 2D perovskites.

Methods

We drop cast some of the remaining supersaturated solution on a glass slide, heated it up to 50 °C with a hotplate and after the solvent was evaporated, crystals with crystal sizes of up to several hundred microns were formed.

$N = 2$ crystals can be easily isolated during the exfoliation (see next section) and the formation of $n = 2$ can be favored by preheating the substrate to 50 °C before drop casting. $(HA)_2PbI_4$ and $(DA)_2PbI_4$ were synthesized by dissolving PbI_2 (100 mg) in HI (800 μl) through heavy stirring and heating the solution to 90 °C.

Perovskite flakes were excited with a 405 nm laser (PicoQuant LDH-D-C-405, PDL 800-D), which was focused down to a near-diffraction limited spot.

A near diffraction limited exciton population was created using a 405 nm laser (PicoQuant LDH-D-C-405, PDL 800-D) and a $\times 100$ oil immersion objective (Nikon CFI Plan Fluor, NA = 1.3).

Conformational Disorder of Organic Cations Tunes the Charge Carrier Mobility in Two-Dimensional Organic-Inorganic Perovskites

DOI: https://doi.org/10.1038/s41467-020-19330-7

This article is licensed under a Creative Commons Attribution 4.0 International License, which permits use, sharing, adaptation, distribution and reproduction in any medium or format, as long as you give appropriate credit to the original author(s) and the source, provide a link to the Creative Commons license, and indicate if changes were made. The images or other third party material in this article are included in the article's Creative Commons license, unless indicated otherwise in a credit line to the material. If material is not included in the article's Creative Commons license and your intended use is not permitted by statutory regulation or exceeds the permitted use, you will need to obtain permission directly from the copyright holder. To view a copy of this license, visit http://creativecommons.org/licenses/by/4.0/.

Originators: Li, C., Yang, J., Su, F. *et al.* Conformational disorder of organic cations tunes the charge carrier mobility in two-dimensional organic-inorganic perovskites. *Nat Commun* 11, 5481 (2020). https://doi.org/10.1038/s41467-020-19330-7

Abstract-Summary

We apply femtosecond broadband sum frequency generation vibrational spectroscopy to investigate the molecular conformation of spacer organic cations in two-dimensional organic-inorganic perovskite films and establish a correlation among the conformation of the organic cations, the charge carrier mobility, and broadband emission.

The gauche defect and local chain distortion of organic cations are the structural origin of the in-plane mobility reduction and broad emission in two-dimensional organic-inorganic perovskites.

This work provides molecular-level understanding of the conformation of organic cations in optimizing the optoelectronic properties of two-dimensional organic-inorganic perovskites.

Introduction

Many studies have shown that chemical tuning of spacer organic cations can provide a simple but effective approach to improve the stability and performance of 2D OIHPs [27–29].

Although important milestones have been achieved through this chemical engineering [30], how the chemical nature of the organic cations within the inorganic framework governs the structural and optoelectronic properties of 2D OIHPs is not fully understood.

Some studies suggest that increasing the length of organic cations can cause structural reorganization, thus leading to the formation of 1D quantum confinement effect [31, 32], while others claim that the charge transport property of 2D OIHP films can be efficiently improved by reducing the interlayer distance and further giving rise to better device performance [33].

We synthesize a series of 2D OIHP films with large organic spacer cations ($[CH_3(CH_2)_{n-1}NH_3]_2PbI_4$, n = 4, 6, 8, 10, 12, 18) using a common procedure [34] and systematically investigate the influence of the molecular structures of cations on their carrier mobility and PL properties.

A remarkable dependence of carrier mobility and broad PL emission on the conformational order of the alkyl chains in organic cations is revealed.

Results

We determine the in-plane charge-carrier mobility of 2D OIHP films using OPTPS experiments.

Out-of-plane charge-carrier mobility also initially increases ($n \leq 6$) and then decreases ($n \geq 6$) as the chain length of the organic cations increases.

It is noticed that these six samples have identical chemical composition of inorganic cage (i.e., the Pb-I layer), the difference in the effective charge-carrier mobility and broadband emission should originate from the packing of the organic cations, which may cause distortion of the inorganic cage.

We then examine molecular conformation of the 2D OIHP thin films at room temperature using femtosecond broadband SFG-VS, to look for correlations between the mobility, broad emission and the structures of organic cations.

This correlation implies that the gauche defect and local chain distortion of organic cations are the structural origin of the in-plane mobility reduction and broad emission in these six 2D OIHP films.

Discussion

We have applied femtosecond broadband SFG-VS to investigate the molecular conformation of spacer organic cations in 2D OIHP films and established a correlation among the conformation of the organic cations, the charge-carrier mobility, and broadband emission.

It is found that both the mobility and broadband emission show strong dependence on the molecular conformational order of organic cations.

It implies that the gauche defect and local chain distortion of organic cations are the structural origin of the in-plane mobility reduction and broad emission in 2D OIHP films.

This work provides physical understanding of the important role of organic cation conformation in optimizing the optoelectronic properties of 2D OIHPs.

Methods

The 2D OIHP films were prepared by spin-coating the precursor solutions on substrates with a spin-coating speed of 1500 rpm at initial 40 s and then 3000 rpm for 20 s. The thicknesses of 9.0, 55, and 144 nm were achieved by spin-coated precursor solutions with concentrations of 0.004, 0.1, and 0.38 M, respectively.

The femtosecond BB-SFG-VS setup was constructed based on a high power regeneration amplifier (Spectra Physics, Spitfire Ace seeded by Mai-Tai SP), which offers 5.0 W output at 800 nm with 13 nm bandwidth, 100 fs pulse duration and 1 kHz repetition rate.

A fraction of a pulse (2.0 W) of the output was used for the excitation of a commercial optical parametric amplifier (TOPAS Prime, Spectra Physics) and collinear difference frequency generation system (with $AgGaS_2$ crystal) to generate tunable broadband infrared pulses (2500–15,000 nm) for the SFG probe.

Phenylalkylammonium Passivation Enables Perovskite Light Emitting Diodes with Record High-Radiance Operational Lifetime: The Chain Length Matters

DOI: https://doi.org/10.1038/s41467-021-20970-6

This article is licensed under a Creative Commons Attribution 4.0 International License, which permits use, sharing, adaptation, distribution and reproduction in any medium or format, as long as you give appropriate credit to the original author(s) and the source, provide a link to the Creative Commons license, and indicate if changes were made. The images or other third party material in this article are included in the article's Creative Commons license, unless indicated otherwise in a credit line to the material. If material is not included in the article's Creative Commons license and your intended use is not permitted by statutory regulation or exceeds the permitted use, you will need to obtain permission directly from the copyright holder. To view a copy of this license, visit http://creativecommons.org/licenses/by/4.0/.

Originators: Guo, Y., Apergi, S., Li, N. *et al.* Phenylalkylammonium passivation enables perovskite light emitting diodes with record high-radiance operational lifetime: the chain length matters. *Nat Commun* 12, 644 (2021). https://doi.org/10.1038/s41467-021-20970-6

Abstract-Summary

We explore surface treatment of perovskite films with phenylalkylammonium iodide molecules of varying alkyl chain lengths.

Combining experimental characterization and theoretical modelling, we show that these molecules stabilize the perovskite through suppression of iodide ion migration.

The stabilization effect is enhanced with increasing chain length due to the stronger binding of the molecules with the perovskite surface, as well as the increased steric hindrance to reconfiguration for accommodating ion migration.

Introduction

Layered two-dimensional (2D) perovskites were incorporated in PeLEDs in consideration of their superior thermal and structural stability compared to the three-dimensional (3D) perovskites [35, 36].

Mixed 2D/3D perovskites have also been introduced to improve the stability of PeLEDs while maintaining high device performance [37, 38], achieving an estimated $T_{50} > 200$ h under initial radiance of about 10 W sr^{-1} m^{-2}.

Li and others provided direct experimental evidence for an ion-migration induced device degradation mechanism and demonstrated largely improved stability of 3D $FAPbI_3$-based PeLEDs by simultaneous incorporation of Cs and Rb cations in the perovskite lattice [39].

Wang and others introduced trifluoroacetate anions in the precursor solution to passivate the surface defects of $CsPbBr_3$ perovskite films and achieved a T_{50} of 250 h at a low luminance of 100 cd m^{-2} (ref.

Developing an effective surface passivation method to suppress the ion-migration and its subsequent device degradation is critical for improving the stability of 3D perovskite-based PeLEDs.

Results

The PL measurements suggest that (1) surface treatment by PMAI (n = 1), PEAI (n = 2), or PPAI (n = 3) can all reduce non-radiative recombination by passivating the surface defects of the perovskite films, corresponding to reduced k_1 values and that (2) surface treatment by PBAI (n = 4) slightly suppresses PL, possibly due to a low anchoring density (i.e., the density of the molecules that are effectively bonded to the perovskite surface) of the passivation molecules and/or the treatment causing formation of new recombination centers that accelerate non-radiative recombination.

In order to examine the effect of the surface treatment on the PeLED performance, we fabricate PeLED devices with a structure of gold (Au)/molybdenum oxide (MoO_3)/poly(9,9-dioctylflu-16orene-co-N-(4-butylphenyl)diphenylamine) (TFB)/ passivation molecule-perovskite/polyethyleneimine (PEIE)/zinc oxide (ZnO)/ indium tin oxide (ITO).

In order to gain an atomistic insight, we perform density functional theory (DFT) calculations to investigate the interactions between the passivation molecules (with alkyl chain n = 1–4) and the perovskite surface.

Having determined the adsorption geometry, we proceed with calculating the binding energy of the passivation molecules with alkyl chain n = 1–4 on the perovskite surface.

Discussion

We explored surface passivation of 3D perovskite films with a series of phenylalkylammonium iodides of varying alkyl chain lengths.

The stabilization effect is enhanced with increasing alkyl chain length, due to the stronger binding of the passivating molecule on the perovskite surface and the increased steric hindrance to reconfiguration required to accommodate I^- migration.

Our work highlights the importance of the chain length effect in suppressing the ion migration, revealing that a very small variation in the surface molecular structure can result in a remarkable improvement in device stability and sheds light on the design of new passivation molecules that could render highly efficient and stable perovskite LEDs.

Methods

The ZnO layer was spun on the substrate at 6000 r.p.m.

PEIE (3.9 mg mL^{-1} in 2-methoxyethanol) layer was spun on ZnO at 5000 r.p.m.

For the geometry optimization and the calculation of the binding energies and the bond orders, 9-layer Pb–I-terminated $FAPbI_3$ slabs were used (cubic unit cell with a fixed lattice parameter of a = 6.36 Å) and a vacuum region of ~30 Å separating the slabs.

The binding energies E_b of the passivation molecules on $FAPbI_3$ were calculated as: where $E_{mol/FAPbI3}$, E_{FAPbI3}, and E_{mol} are the total energies of $FAPbI_3$ with an adsorbed passivation molecule per surface unit cell, the clean $FAPbI_3$ surface, and the $CH_5(CH_2)_nNH_3I$, respectively.

In order to study the effect of the passivation molecules on the migration of I ions, 2 × 2 supercells of the 9-layer slabs were used and optimized by using one k-point (Γ point).

Momentarily Trapped Exciton Polaron in Two-Dimensional Lead Halide Perovskites

DOI: https://doi.org/10.1038/s41467-021-21721-3

This article is licensed under a Creative Commons Attribution 4.0 International License, which permits use, sharing, adaptation, distribution and reproduction in any medium or format, as long as you give appropriate credit to the original author(s) and the source, provide a link to the Creative Commons license, and indicate if changes were made. The images or other third party material in this article are included in the article's Creative Commons license, unless indicated otherwise in a credit line to the material. If material is not included in the article's Creative Commons license and your intended use is not permitted by statutory regulation or exceeds the permitted use, you will need to obtain permission directly from the copyright holder. To view a copy of this license, visit http://creativecommons.org/licenses/by/4.0/.

Originators: Tao, W., Zhang, C., Zhou, Q. *et al.* Momentarily trapped exciton polaron in two-dimensional lead halide perovskites. *Nat Commun* 12, 1400 (2021). https://doi.org/10.1038/s41467-021-21721-3

Abstract-Summary

Compared to their three-dimensional counterparts with large polaron character, how the interplay between long- and short- range exciton-phonon interaction due to polar and soft lattice define the excitons in 2D perovskites is yet to be revealed.

We show quantitatively an intermediate exciton-phonon coupling in 2D $CsPbBr_3$ where exciton polarons are momentarily self-trapped by lattice vibrations.

This study reveals a complex and dynamic picture of exciton polarons in 2D perovskites and emphasizes the importance to regulate exciton-phonon coupling.

Introduction

Exciton–phonon coupling dresses excitons in 2D perovskite with significant polaronic character [40, 41], leading to complex excited-state energetic/configurational landscape [42–44].

Depending on the nature of exciton (carrier)–phonon interaction, exciton (carrier) in lead halide perovskites can exist as exciton polaron (large polaron) which can move freely but dressed with phonon clouds through long-range Frohlich-like interaction, or self-trapped exciton (STE) (small polaron) through short-range Holstein-like interaction [41, 45].

We show that excitons in 2D lead bromide RP perovskites are neither FE polarons nor localized by extrinsic disorders/defects or intrinsic self-trapping effect, but instead, momentarily trapped by lattice fluctuations.

The STE reaches thermal equilibrium with FE polaron in ~0.4 ps at room temperature with a small activation energy of ~34 meV. These results were rationalized with Urbach–Martienssen rule pointing to intermediate exciton–phonon coupling strength and metastable STEs in 2D RP perovskites.

The momentarily trapped exciton polarons in 2D perovskites have strong implications to their optoelectronic applications.

Results

We fit the main peak with a symmetric vogit line shape and denoted it as FE emission since it is sharp and close to the absorption peak and the remaining low-energy tail as localized exciton (LE) emission as it is quite broad and lower in energy.

We also measured PL decay kinetics at different emission energies (from 2.38 to 2.86 eV).

This is also not expected for surface/edge state emission which was observed previously in 2D perovskites and characterized with lower energy and longer PL lifetime compared to FE emission [46].

The energy-independent PL decay kinetics suggests thermal equilibrium between FE and LE at room temperature.

To unambiguously establish the thermal equilibrium and extract the energy difference between FE and LE states, we performed temperature-dependent PL measurements.

This energy difference is only slightly larger than room temperature thermal energy (~25 meV), indicating a significant population of LE state at room temperature.

Discussion

Because of intermediate exciton–phonon coupling in perovskite lattice with dynamic disorder and relatively small lattice configuration change (compared to Holstein-like STE), STE can be directly photoexcited at exponential tail and then establish thermal equilibrium with free state at room temperature, as shown by PL and TA results under low-energy excitation.

This leads to an unique intermediate exciton–phonon coupling regime with momentarily trapped excitons in 2D perovskites, which is different from Fröhlich-like large polaron with dominating long-range interaction and Holstein-like STE with strong short-range interaction.

Previous PL results on $(BA)_2PbBr_4$ single crystal show a more prominent low-energy tail [47], suggesting more STE character with stronger exciton–phonon interaction for lead bromide perovskite.

We show that the asymmetric PL spectrum with low-energy exponential tail is intrinsic and originates from momentarily trapped exciton polarons with intermediate exciton–phonon interaction strength.

Methods

To a solution of 1.25 mL of 1-octadecene (ODE), 0.125 mL of oleylamine (OLAM) and 0.125 mL of oleic acid (OA), 7 μL of hydrobromic acid (HBr, 48 wt.% in H_2O) and 0.1 mL of Cs-oleic acid precursor (0.1 M, 0.35 g of Cs_2CO_3 degassed in 20 mL of ODE and 1.25 mL of OA at 150 °C) was swiftly injected 0.2 mL of $PbBr_2$ precursor (0.4 M, 735 mg of $PbBr_2$ in 5 mL of DMF) under stirring.

The solution turned turbid white within several seconds, and after 10 s, 5 mL of acetone was swiftly added to quench the reaction.

After 1-min stirring, the NPs were collected by holding the supernatant after centrifugation at 8000 r.p.m.

Temperature-dependent PL and TRPL measurements were taken on a home-built microscope setup.

Light-Activated Interlayer Contraction in Two-Dimensional Perovskites for High-Efficiency Solar Cells

DOI: https://doi.org/10.1038/s41565-021-01010-2

Originators: Li, W., Sidhik, S., Traore, B. *et al.* Light-activated interlayer contraction in two-dimensional perovskites for high-efficiency solar cells. *Nat. Nanotechnol.* 17, 45–52 (2022). https://doi.org/10.1038/s41565-021-01010-2

Abstract-Summary

Correlated charge transport, structural and photovoltaic measurements confirm that the onset of the light-induced contraction is synchronized to a threefold increase in carrier mobility and conductivity, which is consistent with an increase in the electronic band dispersion predicted by first-principles calculations.

Flux-dependent space-charge-limited current measurement reveals that light-induced interlayer contraction activates interlayer charge transport.

The enhanced charge transport boosts the photovoltaic efficiency of two-dimensional perovskite solar cells up to 18.3% by increasing the device's fill factor and open-circuit voltage.

Main

The composition of the organic moieties and the perovskite layer thickness (expressed by the n value in the chemical formula $L_2A_{n-1}B_nX_{3n+1}$ where L is the organic monovalent spacer; A is methylammonium (MA), formamidinium (FA) or Cs; B is a divalent metal such as Pb or Sn; and X is a halide: I, Br or Cl) define the crystal phase, structural, mechanical, electronic and optical properties of resulting 2D perovskites [30, 48–56].

The stacking of the perovskite layers and the use of the short cation (4AMP) in the DJ structure impose a perfectly eclipsed alignment on the layer stacking, resulting in weak but notable electronic coupling between the iodine atoms directly facing each other across the organic interlayer [57, 58].

Tailoring interlayer interactions by using suitable organic cations or external mechanical stress has been predicted to enhance the electronic coupling between perovskite layers across the organic spacer and significantly enhance the charge transport in 2D hybrid perovskites [48, 59–64].

Structural characterization of 2D perovskite under light

These relative changes in the lattice parameter values over time, which can be interpreted as the normal strain induced in the lattice by light, reveal that the light-induced lattice contraction effect is significantly stronger in the direction of the stacking axis (out-of-plane) as compared to the in-plane direction.

The continuous and reversible light-induced lattice contraction of the original perovskite phase is an intrinsic phenomenon observed for all types of samples and is a bulk property.

We observed that both the ACI $n = 2$ and $n = 3$ showed light-induced structural changes similar to the DJ $n = 3$.

These measurements suggest that a correlation exists between the different structural phases of 2D perovskite and the effect of light-induced contraction.

Mechanism of light-activated interlayer contraction

The blue shift in the bonding states indicates the loss of electrons (or oxidation) of the iodide ion.

The more positively charged iodides across the interlayer can undergo an oxidative coupling, which results in the out-of-plane contraction and a decrease in the effective interlayer distance.

These XPS trend measurements clearly imply that light-induced out-of-plane contraction is correlated to the interlayer iodide interactions in DJ and ACI 2D perovskites, in which the interlayer distance is the shortest.

Because of the proximity of the adjacent slabs enabled by the very short spacer cations, some holes can get trapped through the enhanced I–I weakly bonding interactions across the slabs, resulting in contraction along the out-of-plane direction.

A key requirement for this mechanism to be activated is that the I–I distance across the slabs must be short; therefore, this effect of light-activated interlayer contraction is observed only in DJ and ACI perovskites.

In-situ transport and device measurements

Motivated by the DFT band-structure calculations, which predict an increase in band dispersion upon light-induced contraction in the out-of-plane direction, the next logical step was to quantify the impact of contraction on the electronic transport properties.

These results suggest that the charge transport is significantly enhanced within the short time of 10 minutes, even though the effect of light-induced contraction had just initiated.

To gain deeper insight into the correlation between the light-induced contraction and the enhancement in charge transport, we performed light-intensity-dependent GIWAXS and SCLC measurements, described below.

Similar to the single crystals and powder samples, thin films of DJ $n = 3$ also exhibit a light-induced anisotropic compressive strain resulting in the contraction of the effective interlayer distance.

These correlated structure, transport and device data clearly imply that the drastic improvement in the device performance is driven predominantly by the reduction of the effective interlayer distance resulting in the increase in carrier mobility, especially along the stacking direction.

Solar cell performance

The overall increase in the power conversion efficiency of DJ and ACI 2D perovskite solar cells arises from a combined increase in the V_{OC} and FF.

In agreement with the DJ powder and single-crystal results, we observed a decrease to the original solar cell efficiency after 23 minutes of resting the device in the dark.

The largest increase for the champion DJ device resulted in the increase in the power conversion efficiency from 15.6% to 18.3%.

Our results indicate that the light-induced out-of-plane lattice contraction, which results from the enhanced I–I bonding across the interlayers in DJ and ACI perovskite thin films, lowers the depth of the potential barriers, thus decreasing the number of trapped carriers and thereby reducing the recombination rate.

Conclusion

They demonstrate that by selecting the appropriate interlayer cations and by leveraging the acute sensitivity of 2D hybrid perovskites to external stimuli, the bandgap discontinuity in the out-of-plane direction can be surmounted, and charge transport properties closer to those of 3D perovskites can be realized.

These results pave the path for understanding and tailoring new light–matter interactions, similar to those explored in 2D transition-metal dichalcogenides, which arise when excitations are coupled across the layered perovskite layers separated by short organic cations [65, 66].

Methods

For the in-situ light-illuminated crystal GIWAXS measurements, we installed a Newport Class ABB Solar Simulator on top of each beamline chamber (schematic (1) staging chamber for beamline 11-BM) and illuminated with 1 Sun (AM 1.5G) irradiance.

Before and after light illumination, XPS was performed on large millimetre-size DJ and RP n = 3 single-crystal perovskites and films grown on a glass substrate using the procedure mentioned in the first section of the Methods.

To minimize the latter effect, the cells were doubled along the a direction for both n = 2 and 3 of the ACI compounds resulting in the a lattice parameter being about 12.8 Å. Due to the molecular positions in the X-ray diffraction data, which are subject to ambiguity because of their dynamic disorder, especially at room temperature, we rotated every second MA molecule to 180° in each plane of the perovskite layer for the ACI compounds [67].

Electron-Phonon Interactions in Halide Perovskites

DOI: https://doi.org/10.1038/s41427-022-00394-4

This article is licensed under a Creative Commons Attribution 4.0 International License, which permits use, sharing, adaptation, distribution and reproduction in any medium or format, as long as you give appropriate credit to the original author(s) and the source, provide a link to the Creative Commons license, and indicate if changes were made. The images or other third party material in this article are included in the article's Creative Commons license, unless indicated otherwise in a credit line to the material. If material is not included in the article's Creative Commons license and your intended use is not permitted by statutory regulation or exceeds the permitted use, you will need to obtain permission directly from the copyright holder. To view a copy of this license, visit http://creativecommons.org/licenses/by/4.0/.

Originators: Yamada, Y., Kanemitsu, Y. Electron-phonon interactions in halide perovskites. *NPG Asia Mater* 14, 48 (2022). https://doi.org/10.1038/s41427-022-00394-4

Abstract-Summary

Strong electron-phonon interactions are frequently considered the origin of the unique electrical and optical properties of lead halide perovskites.

Electron-phonon interactions induce the formation of a polaron, which is a charge carrier dressed with a phonon cloud.

While the Fröhlich interaction plays an essential role in ionic crystals, we also need to consider the strong phonon anharmonicity of halide perovskites that may lead to the formation of an unconventional polaron.

Article, we discuss the uniqueness of perovskite semiconductors from the viewpoint of electron-phonon interactions.

Introduction

Since the effective masses of the electron and the hole in lead halide perovskites are approximately 0.1 m_0 (and thus are not much different from conventional inorganic semiconductors), the polaron effect has frequently been considered the origin of the low mobility.

Several theoretical investigations based on the Fröhlich mechanism have suggested that the low optical phonon energies in perovskites are the reason for their low carrier mobilities; [68–70] a small LO phonon energy implies that many phonons are excited at room temperature, and therefore, strong scattering occurs.

While electron-phonon interactions in metal halide perovskites can result in a reduction in mobility or induce ion migration and thus partially constitute a harmful factor in device applications, they also lead to slow relaxation of hot carriers as well as anti-Stokes PL, and the utilization of these unique phenomena in new devices is expected.

Ultrafast Relaxation of Lattice Distortion in Two-Dimensional Perovskites

DOI: https://doi.org/10.1038/s41567-022-01903-6

Originators: Zhang, H., Li, W., Essman, J. *et al.* Ultrafast relaxation of lattice distortion in two-dimensional perovskites. *Nat. Phys.* 19, 545–550 (2023). https://doi.org/10.1038/s41567-022-01903-6

Abstract-Summary

Direct visualization of ultrafast coupling between charge carriers and lattice degrees of freedom in photoexcited semiconductors has remained a long-standing challenge and is critical for understanding the light-induced physical behaviour of materials under extreme non-equilibrium conditions.

Our analysis reveals a light-induced ultrafast reduction in antiferro-distortion resulting from a strong interaction between the electron–hole plasma and perovskite lattice, which induces an in-plane octahedra rotation towards a more symmetric phase.

Correlated ultrafast spectroscopy performed at the same carrier density as ultrafast electron diffraction reveals that the creation of a dense electron–hole plasma triggers the relaxation of lattice distortion at shorter timescales by modulating the crystal cohesive energy.

Main

These measurements indicate that the presence of an organic cation in close proximity to the inorganic perovskite lattice strongly modulates the nature of electron–phonon interactions [18, 26, 40, 71–76], and suggest that electron–phonon scattering in 2D perovskites occurs via local dynamic disorder [64].

There exists no direct ultrafast structural measurement of 2D perovskites, which elucidates the underlying carrier–lattice interaction mechanisms on optical excitation.

We report a direct measurement of the structural dynamics in 2D perovskites obtained by monitoring the change in femtosecond-scale ultrafast electron diffraction (UED) after optical excitation.

This technique enables the picosecond time-resolved structural evolution of 2D perovskites by tracking the changes in diffraction pattern, thus providing a direct visualization of lattice response after photodoping.

We show that the mechanism and dynamics of the interaction between the charge carriers and lattice are acutely tunable and sensitive to the initial lattice distortion of 2D perovskites dictated by the choice of the organic spacer layers.

UED experiments on 2D perovskites

Monitoring the characteristics of the diffraction peaks (or Bragg peaks) on the image allows us to directly probe the dynamic changes in the lattice structure after light excitation, revealing the carrier–lattice interactions during energy relaxation (cooling) to the band edge of 2D perovskite semiconductors.

A first visual inspection of these data reveals an anisotropic response with respect to the in-plane directions in the 2D perovskite lattice as well as a monotonic decrease in the absolute intensity response with respect to the magnitude of scattering vector $|q|$.

We note that opposite to our results, conventional well-studied bulk semiconductors (such as Si and GaAs), as well as inorganic 2D materials (for example, transition metal dichalcogenides such as MoS_2 and $MoSe_2$), exhibit—in most cases—a decrease in Bragg peak intensities after photoexcitation [77–79].

The increase in the intensities of specific Bragg peaks, therefore, counteracts the Debye–Waller effect at shorter timescales and reduces the lattice distortion (enhanced order) in specific directions.

Mechanism of transient lattice ordering

We attribute this deviation to the combination of both light-induced reduction in antiferro-distortion (ordering) and Debye–Waller response.

The Bragg peaks with positively changing intensity, for which the light-induced lattice ordering mechanism dominates ((400) and (040)), exhibit a very fast time constant of 1 ps or shorter, whereas the peaks with a negative intensity response yield a longer time constant of the order of 5 ps.

Similar time constants have also been reported in 3D perovskites and have been attributed to the Debye–Waller-like response originating from the rotational disorder of the iodine atoms in the perovskite octahedra, exclusively leading to a decrease in peak intensities [80].

We propose that the light-induced lattice dynamics can be attributed to the reduction in in-plane lattice antiferro-distortion at ultrafast timescales ($\tau \leq 1$ ps) and to a Debye–Waller-like response at longer timescales ($\tau \approx 5$ ps).

Lattice transients triggered by dense electron–hole plasma

We conclude that the observed energy transfer at shorter timescales in the UED experiment is rather related to the coupling of a dense carrier plasma to the lattice, instead of exciton polaronic couplings reported in optical studies on n = 1 2D perovskites at low temperatures and low excitation fluences [40].

In the case of a dense electron–hole plasma, strong transient lattice distortions different from Debye–Waller effects can be induced in semiconductors, generating supersonically propagating strain pulses [81, 82], as well as more complex lattice reorganizations such as the suppression of ferroelectric instability in $KTaO_3$ (ref 89), and the activation of the antiferro-distorsive rotation of octahedra in $SrTiO_3$ (ref 90).

This last mechanism of plasma-induced modulation of the interatomic force must be considered because halide perovskites are strongly ionic materials like $SrTiO_3$.

The resulting modification of the crystal cohesive energy manifests as strong lattice reorganization with the antiferro-distortive rotation of octahedra in 2D DJ perovskites.

A more classical decay of the plasma energy occurs through electron–phonon interactions and observed in structural dynamics associated with a Debye–Waller effect.

Organic cations alter carrier–lattice interactions

To understand if the mechanism of light-induced transformation is generally applicable to other 2D perovskites with different organic cations and structural phases, we perform UED experiments on three other hybrid 2D perovskites: 4AMP-$MA_2Pb_3I_{10}$ (DJ n = 3) exhibiting the same composition as DJ n = 2 but a thicker perovskite layer, $BA_2MAPb_2I_7$ (RP n = 2) with the same perovskite layer as DJ n = 2 but different spacer organic cations (with butylammonium (BA)), and $BA_2MA_3Pb_4I_{13}$ (RP n = 4) with a different perovskite layer and spacer organic cations.

To quantify the different behaviours between the DJ- and RP-phase 2D perovskites, we evaluated the differential intensity and time constant of the relevant Bragg peaks.

These results demonstrate that there is a fundamental difference in the mechanism of carrier–lattice coupling and structural dynamics between the DJ and RP 2D perovskite phases, which implies that the competing mechanism of ultrafast lattice ordering during an antiferro-distortive phase transformation is mitigated in the RP phase.

Methods

Normal-incident pulses were diffracted by the sample in transmission geometry on the transmission electron microscopy window, and the diffracted intensities were captured by an electron-multiplying charge-coupled device.

The Bragg peaks of the measured samples (DJ n = 2, DJ n = 3, RP n = 2 and RP n = 4) were indexed by comparing the diffraction patterns with the simulated diffraction images calculated via SingleCrystal software (version 3.1, http://crysta

lmaker.com/singlecrystal/), using structure files (crystallographic information files (.cif)) from another work [57].

The time traces of Bragg peaks were globally fitted by a single exponential curve:where Δ stands for intensity change (%), τ denotes the time constant (ps) and t_0 is the time-zero offset (calibrated by the response of bismuth crystal and globally fixed).

2D perovskites were measured in the ambient air condition with the DJ n = 2 sample sealed with 50-nm-thick polymethyl methacrylate on top.

Bibliography

1. Yu, P.Y., Cardona, M.: Fundamentals of Semiconductors. Graduate Texts in Physics Springer-Verlag (2010)
2. Stillman, G., Wolfe, C., Dimmock, J.: Hall coefficient factor for polar mode scattering in n-type GaAs. J. Phys. Chem. Solids **31**, 1199–1204 (1970)
3. Milot, R.L., Eperon, G.E., Snaith, H.J., Johnston, M.B., Herz, L.M.: Temperature-dependent charge-carrier dynamics in $CH_3NH_3PbI_3$ perovskite thin films. Adv. Funct. Mater. **25**, 6218–6227 (2015)
4. Xiao, M., et al.: A fast deposition-crystallization procedure for highly efficient lead iodide perovskite thin-film solar cells. Angew. Chem. Int. Ed. **126**, 10056–10061 (2014)
5. Ong, W.-L., Rupich, S.M., Talapin, D.V., McGaughey, A.J.H., Malen, J.A.: Surface chemistry mediates thermal transport in three-dimensional nanocrystal arrays. Nat. Mater. **12**, 410–415 (2013)
6. Loynachan, C.N., et al.: Targeted magnetic nanoparticles for remote magnetothermal disruption of amyloid-β aggregates. Adv. Healthcare Mater. **4**, 2100–2109 (2015)
7. Love, J.C., Estroff, L.A., Kriebel, J.K., Nuzzo, R.G., Whitesides, G.M.: Self-assembled monolayers of thiolates on metals as a form of nanotechnology. Chem. Rev. **105**, 1103–1170 (2005)
8. Losego, M.D., Grady, M.E., Sottos, N.R., Cahill, D.G., Braun, P.V.: Effects of chemical bonding on heat transport across interfaces. Nat. Mater. **11**, 502–506 (2012)
9. Chiritescu, C., et al.: Ultralow thermal conductivity in disordered, layered WSe_2 crystals. Science **315**, 351–353 (2007)
10. Ravichandran, J., et al.: Crossover from incoherent to coherent phonon scattering in epitaxial oxide superlattices. Nat. Mater. **13**, 168–172 (2014)
11. Simkin, M.V., Mahan, G.D.: Minimum thermal conductivity of superlattices. Phys. Rev. Lett. **84**, 927–930 (2000)
12. Maldovan, M.: Phonon wave interference and thermal bandgap materials. Nat. Mater. **14**, 667–674 (2015)
13. Li, D., McGaughey, A.J.H.: Phonon dynamics at surfaces and interfaces and its implications in energy transport in nanostructured materials—an opinion paper. Nanoscale Microsc. Thermophys. Eng. **19**, 166–182 (2015)
14. Ong, W.-L., et al.: Orientational order controls crystalline and amorphous thermal transport in superatomic crystals. Nat. Mater. **16**, 83–88 (2017)
15. Ivanovska, T., et al.: Vibrational response of methylammonium lead iodide: from cation dynamics to phonon–phonon interactions. Chemsuschem **9**, 2994–3004 (2016)
16. Yaffe, O., et al.: Excitons in ultrathin organic–inorganic perovskite crystals. Phys. Rev. B **92**, 045414 (2015)
17. Srimath Kandada, A.R., Petrozza, A.: Photophysics of hybrid lead halide perovskites: the role of microstructure. Acc. Chem. Res. **49**, 536–544 (2016)

18. Miyata, K., et al.: Large polarons in lead halide perovskites. Sci. Adv. **3**, e1701217 (2017)
19. Neutzner, S., et al.: Exciton-polaron spectral structures in two-dimensional hybrid lead-halide perovskites. Phys. Rev. Mater. **2**, 064605 (2018)
20. Yang, S., Fu, W., Zhang, Z., Chen, H., Li, C.Z.: Recent advances in perovskite solar cells: efficiency, stability and lead-free perovskite. J. Mater. Chem. A **5**, 11462–11482 (2017)
21. Smith, I.C., Hoke, E.T., Solis-Ibarra, D., McGehee, M.D., Karunadasa, H.I.: A layered hybrid perovskite solar-cell absorber with enhanced moisture stability. Angew. Chem. Int. Ed. **53**(42), 11232–11235 (2014)
22. Quan, L.N., Yuan, M., Comin, R., Voznyy, O., Beauregard, E.M., Hoogland, S., Buin, A., Kirmani, A.R., Zhao, K., Amassian, A., Kim, D.H., Sargent, E.H.: Ligand-stabilized reduced-dimensionality perovskites. J. Am. Chem. Soc. **138**(8), 2649–2655 (2016). https://doi.org/10.1021/jacs.5b11740
23. Fu, Q., et al.: Recent progress on the long-term stability of perovskite solar cells. Adv. Sci. **5** (2018)
24. Ortiz-Cervantes, C., Carmona-Monroy, P., Solis-Ibarra, D.: Two-dimensional halide perovskites in solar cells: 2D or not 2D? Chemsuschem **12**, 1560–1575 (2019)
25. Blancon, J.-C., et al.: Extremely efficient internal exciton dissociation through edge states in layered 2D perovskites. Science **355**, 1288–1292 (2017)
26. Akselrod, G.M., et al.: Visualization of exciton transport in ordered and disordered molecular solids. Nat. Commun. **5**, 3646 (2014)
27. Gong, X., et al.: Electron–phonon interaction in efficient perovskite blue emitters. Nat. Mater. **17**, 550–556 (2018)
28. Hu, J., Oswald, I.W.H., Stuard, S.J., Nahid, M.M., Zhou, N., Williams, O.F., Guo, Z., Yan, L., Hu, H., Chen, Z., Xiao, X., Lin, Y., Yang, Z., Huang, J., Moran, A.M., Ade, H., Neilson, J.R., You, W.: Synthetic control over orientational degeneracy of spacer cations enhances solar cell efficiency in two-dimensional perovskites. Nat. Commun. **10**(1), 1276 (2019). https://doi.org/10.1038/s41467-019-08980-x
29. Manser, J.S., Christians, J.A., Kamat, P.V.: Intriguing optoelectronic properties of metal halide perovskites. Chem. Rev. **116**, 12956–13008 (2016)
30. Straus, D.B., Kagan, C.R.: Electrons, excitons, and phonons in two-dimensional hybrid perovskites: connecting structural, optical, and electronic properties. J. Phys. Chem. Lett. **9**(6), 1434–1447 (2018). https://doi.org/10.1021/acs.jpclett.8b00201
31. Mao, L., Stoumpos, C.C., Kanatzidis, M.G.: Two-dimensional hybrid halide perovskites: principles and promises. J. Am. Chem. Soc. **141**, 1171–1190 (2019)
32. Gan, L., Li, J., Fang, Z.S., He, H.P., Ye, Z.Z.: Effects of organic cation length on exciton recombination in two-dimensional layered lead iodide hybrid perovskite crystals. J. Phys. Chem. Lett. **8**, 5177–5183 (2017)
33. Kamminga, M.E., et al.: Confinement effects in low-dimensional lead iodide perovskite hybrids. Chem. Mater. **28**, 4554–4562 (2016)
34. Yu, M.T., et al.: Control of barrier width in perovskite multiple quantum wells for high performance green light–emitting diodes. Adv. Optical Mater. **7**, 1801575 (2019)
35. Ni, L.M., et al.: Real-time observation of exciton-phonon coupling dynamics in self-assembled hybrid perovskite quantum wells. ACS Nano **11**, 10834–10843 (2017)
36. Deng, W., et al.: 2D Ruddlesden-Popper perovskite nanoplate based deep-blue light-emitting diodes for light communication. Adv. Funct. Mater. **29**, 1–9 (2019)
37. Shang, Y., et al.: Highly stable hybrid perovskite light-emitting diodes based on Dion-Jacobson structure. Sci. Adv. **5**, eaaw8072 (2019)
38. Han, T.H., et al.: Surface-2D/Bulk-3D heterophased perovskite nanograins for long-term-stable light-emitting diodes. Adv. Mater. **32**, 1–10 (2020)
39. Li, C., et al.: Understanding the improvement in the stability of a self-assembled multiple-quantum well perovskite light-emitting diode. J. Phys. Chem. Lett. **10**, 6857–6864 (2019)
40. Li, N., et al.: Stabilizing perovskite light-emitting diodes by incorporation of binary alkali cations. Adv. Mater. **32**, 1907786 (2020)

41. Thouin, F., et al.: Phonon coherences reveal the polaronic character of excitons in two-dimensional lead halide perovskites. Nat. Mater. **18**, 349–356 (2019)
42. Srimath Kandada, A.R., Silva, C.: Exciton polarons in two-dimensional hybrid metal-halide perovskites. J. Phys. Chem. Lett. 11, 3173–3184 (2020)
43. Thouin, F., et al.: Electron–phonon couplings inherent in polarons drive exciton dynamics in two-dimensional metal-halide perovskites. Chem. Mater. **31**, 7085–7091 (2019)
44. Neutzner, S., et al.: Exciton-polaron spectral structures in two-dimensional hybrid lead-halide perovskites. Phys. Rev. Mater. **2**, 6 (2018)
45. Nan, G., Beljonne, D., Zhang, X., Quarti, C.: Organic cations protect methylammonium lead iodide perovskites against small exciton-polaron formation. J. Phys. Chem. Lett. **11**, 2983–2991 (2020)
46. Ghosh, D., Welch, E., Neukirch, A.J., Zakhidov, A., Tretiak, S.: Polarons in halide perovskites: a perspective. J. Phys. Chem. Lett. **11**, 3271–3286 (2020)
47. Shi, E., et al.: Extrinsic and dynamic edge states of two-dimensional lead halide perovskites. ACS Nano **13**, 1635–1644 (2019)
48. Chen, H., et al.: Structural and spectral dynamics of single-crystalline Ruddlesden-Popper phase halide perovskite blue light-emitting diodes. Sci. Adv. **6**, eaay4045 (2020)
49. Billing, D.G., Lemmerer, A.: Synthesis, characterization and phase transitions of the inorganic–organic layered perovskite-type hybrids [$(CnH_{2n}{+}1NH_3)_2PbI_4$] (n = 12, 14, 16 and 18). New J. Chem. **32**, 1736–1746 (2008)
50. Billing, D.G., Lemmerer, A.: Synthesis and crystal structures of inorganic–organic hybrids incorporating an aromatic amine with a chiral functional group. CrystEngComm **8**, 686–695 (2006)
51. Mitzi, D.B.: In: Karlin, K.D. (eds.) Progress in Inorganic Chemistry, vol. 48, pp. 1–121. Wiley (1999). https://doi.org/10.1002/9780470166499.ch1
52. Tu, Q., et al.: Out-of-plane mechanical properties of 2D hybrid organic–inorganic perovskites by nanoindentation. ACS Appl. Mater. Interfaces **10**, 22167–22173 (2018)
53. Reyes-Martinez, M.A., et al.: Unraveling the elastic properties of (quasi) two-dimensional hybrid perovskites: a joint experimental and theoretical study. ACS Appl. Mater. Interfaces **12**, 17881–17892 (2020)
54. Katan, C., Mercier, N., Even, J.: Quantum and dielectric confinement effects in lower-dimensional hybrid perovskite semiconductors. Chem. Rev. **119**, 3140–3192 (2019)
55. Spanopoulos, I., et al.: Uniaxial expansion of the 2D Ruddlesden-Popper perovskite family for improved environmental stability. J. Am. Chem. Soc. **141**, 5518–5534 (2019)
56. Gompel, W.T.M.V., et al.: Towards 2D layered hybrid perovskites with enhanced functionality: introducing charge-transfer complexes via self-assembly. Chem. Commun. **55**, 2481–2484 (2019)
57. Ahn, J., et al.: A new class of chiral semiconductors: chiral-organic-molecule-incorporating organic–inorganic hybrid perovskites. Mater. Horiz. **4**, 851–856 (2017)
58. Mao, L., et al.: Hybrid Dion-Jacobson 2D lead iodide perovskites. J. Am. Chem. Soc. **140**, 3775–3783 (2018)
59. Soe, C.M.M., et al.: New type of 2D perovskites with alternating cations in the interlayer space, $(C(NH_2)_3)(CH_3NH_3)nPbnI_{3n+1}$: structure, properties, and photovoltaic performance. J. Am. Chem. Soc. **139**, 16297–16309 (2017)
60. Soe, C.M.M., et al.: Structural and thermodynamic limits of layer thickness in 2D halide perovskites. Proc. Natl. Acad. Sci. U.S.A. **116**, 58–66 (2019)
61. Leng, K., et al.: Molecularly thin two-dimensional hybrid perovskites with tunable optoelectronic properties due to reversible surface relaxation. Nat. Mater. **17**, 908–914 (2018)
62. Liu, G., et al.: Isothermal pressure-derived metastable states in 2D hybrid perovskites showing enduring bandgap narrowing. Proc. Natl. Acad. Sci. U.S.A. **115**, 8076–8081 (2018)
63. Liu, S., et al.: Manipulating efficient light emission in two-dimensional perovskite crystals by pressure-induced anisotropic deformation. Sci. Adv. **5**, eaav9445 (2019)
64. Yu, S., et al.: Nonconfinement structure revealed in Dion-Jacobson type quasi-2D perovskite expedites interlayer charge transport. Small **15**, 1905081 (2019)

65. Blancon, J.-C., Even, J., Stoumpos, C.C., Kanatzidis, M.G., Mohite, A.D.: Semiconductor physics of organic–inorganic 2D halide perovskites. Nat. Nanotechnol. **15**, 969–985 (2020)
66. Britnell, L., et al.: Strong light-matter Interactions in heterostructures of atomically thin films. Science **340**, 1311–1314 (2013)
67. Xia, F., Wang, H., Xiao, D., Dubey, M., Ramasubramaniam, A.: Two-dimensional material nanophotonics. Nat. Photon. **8**, 899–907 (2014)
68. Long, G., et al.: Spin control in reduced-dimensional chiral perovskites. Nat. Photon. **12**, 528–533 (2018)
69. Sendner, M., et al.: Optical phonons in methylammonium lead halide perovskites and implications for charge transport. Mater. Horiz. **3**, 613–620 (2016)
70. Frost, J.M.: Calculating polaron mobility in halide perovskites. Phys. Rev. B **96**, 195202 (2017)
71. Poncé, S., Schlipf, M., Giustino, F.: Origin of low carrier mobilities in halide perovskites. ACS Energy Lett. **4**, 456–463 (2019)
72. Quan, L.N., et al.: Vibrational relaxation dynamics in layered perovskite quantum wells. Proc. Natl. Acad. Sci. U.S.A. **118**, e2104425118 (2021)
73. Guo, Z., Wu, X., Zhu, T., Zhu, X., Huang, L.: Electron–phonon scattering in atomically thin 2D perovskites. ACS Nano **10**, 9992–9998 (2016)
74. Jia, X., et al.: Observation of enhanced hot phonon bottleneck effect in 2D perovskites. Appl. Phys. Lett. **112**, 143903 (2018)
75. Straus, D.B., et al.: Direct observation of electron–phonon coupling and slow vibrational relaxation in organic–inorganic hybrid perovskites. J. Am. Chem. Soc. **138**, 13798–13801 (2016)
76. Ni, L., et al.: Real-time observation of exciton–phonon coupling dynamics in self-assembled hybrid perovskite quantum wells. ACS Nano **11**, 10834–10843 (2017)
77. Guo, P., et al.: Direct observation of bandgap oscillations induced by optical phonons in hybrid lead iodide perovskites. Adv. Funct. Mater. **30**, 1907982 (2020)
78. Harb, M., et al.: Carrier relaxation and lattice heating dynamics in silicon revealed by femtosecond electron diffraction. J. Phys. Chem. B **110**, 25308–25313 (2006)
79. Lin, M.-F., et al.: Ultrafast non-radiative dynamics of atomically thin $MoSe_2$. Nat. Commun. **8**, 1745 (2017)
80. Mannebach, E.M., et al.: Dynamic structural response and deformations of monolayer MoS_2 visualized by femtosecond electron diffraction. Nano Lett. **15**, 6889–6895 (2015)
81. Wu, X., et al.: Light-induced picosecond rotational disordering of the inorganic sublattice in hybrid perovskites. Sci. Adv. **3**, e1602388 (2017)
82. DeCamp, M.F., et al.: Transient strain driven by a dense electron-hole plasma. Phys. Rev. Lett. **91**, 165502 (2003)
83. Young, E.S.K., Akimov, A.V., Campion, R.P., Kent, A.J., Gusev, V.: Picosecond strain pulses generated by a supersonically expanding electron-hole plasma in GaAs. Phys. Rev. B **86**, 155207 (2012)
84. Krapivin, V., et al.: Ultrafast suppression of the ferroelectric instability in $KTaO_3$. Phys. Rev. Lett. **129**, 127601 (2022)
85. Porer, M., et al.: Ultrafast relaxation dynamics of the antiferrodistortive phase in Ca doped $SrTiO_3$. Phys. Rev. Lett. **121**, 055701 (2018)

Chapter 4
Applications I: Photovoltaics and Optical Detectors

Introduction by Editor

Optical applications have been and remain a huge motivation for metal halide perovskite research, including both 3D MHPs and 2D MHPs. The ease of fabrication is an obvious advantage, as well as the tunability through structural as well as chemical alterations. This section comprises 14 papers, including 12 which focus on photovoltaic applications and two which discuss 2D MHPs used as optical detectors. The papers are presented chronologically within each grouping.

With regards to photovoltaics, historically the large exciton binding energy of 2D MHPs has provided a disadvantage in optical absorption applications, since to form a usable current the exciton must be split into free electron and hole charge carriers. However, 2D MHPs have been shown to be much more stable than 3D MHPs, as the organic molecules in the 2D materials can block ion migration and protect against water-induced degradation. Consequently the bulk of these research papers deal with attempts to get the best of both worlds—high performance as well as long practical lifetimes.

One route to getting the best of both of the 2D and 3D MHP worlds involves a literal merger of 2D and 3D materials—for example, a 3D MHP material can be coated with a 2D MHP, or 2D platelets can be interspersed between 3D perovskite grains. Another route is to control the *n*-number of the 2D material, which as has been mentioned is the number of metal halide monolayers that comprise each inorganic layer. An increased n number (n = a few) can potentially strike a nice balance between photovoltaic performance (reduced exciton binding energy compared to strictly 2D) and stability (organic layers present to block ion migration compared to 3D). Tuning the n number of the 2D material in a 2D/3D hybrid can also potentially assist with better charge transport across the interface. To further counter possible ion migration, one of the papers presented here includes a cross-linked polymer as an additional

© The Author(s), under exclusive license to Springer Nature Singapore Pte Ltd. 2024

J. S. Colton and K. R. Hansen (eds.), *Two-Dimensional Metal Halide Perovskites*,
https://doi.org/10.1007/978-981-99-7830-4_4

layer between the 2D and 3D components of the hybrid MHP device. Finally, due to the combined effects of quantum and dielectric confinement (see Chap. 2), 2D MHs have larger bandgaps than their 3D counterparts; therefore the n numbers can be used to tune bandgaps between n = 1 and n = ∞ limits, giving yet one more avenue of tunability for 2D MHPs.

However, increasing the n number in 2D MHPs typically leads a random distribution of well widths centered on the targeted value; i.e., it can be difficult to create phase-pure material with higher n numbers. This can cause its own set of issues, for example with the inhomogeneous energy landscape introducing extra energy loss, creating band tail states, and inhibiting minority carrier transport. Therefore some research efforts are looking for ways to increase homogeneity, obtaining perovskite phases with identical values for n. Another route is to be more tactical with the n number control; for example it can be advantageous to form small n number layers close to the substrate and large n number layers father away.

In all of this, the organic cations give a degree of tunability for 2D MHPs that is simply not present in 3D MHPs. Molecules can be selected to tune bandgaps and other electronic properties, and can also help increase transport. Hydrophobic cations can be added to surfaces to protect against water. The organic cations can also tune the orientation of the planes to help control transport, e.g. planes aligned parallel to the substrate vs perpendicularly as has been mentioned before in Chap. 2.

In addition to photovoltaics, photodetectors comprise another promising set of optical absorption-related applications. The strong excitonic effects give rise to a strong absorption, which means that 2D MHP-based detectors can be made much thinner than 3D MHPs. The dominating excitonic absorption peak occurs at a definite wavelength and is therefore is both narrowband and highly tunable. Finally, the thin layer combined with the typical ease of MHP fabrication also can lead towards interesting geometries, such as (presented here) coating a hemisphere.

Machine generated summaries

Disclaimer: The auto-summaries have been generated by a recursive clustering algorithm via the Dimensions Auto-summarizer by Digital Science handled by Subject Matter Experts and the external editor. The editor of this book selected which SN content should be auto-summarized and decided its order of appearance. Please be aware that the auto-summaries consist of original sentences, but are not representative of its original paper, since we do not show the full length of the publication. Please note that only published SN content is represented here, and that machine-generated books are still at an experimental stage.

Machine generated keywords: pce, psc, cell, vertical, degradation, stabilize, perovskite solar, ammonium, solar, solar cell, distribution, bulky, surface, heterostructure, spacer

One-Year Stable Perovskite Solar Cells by 2D/3D Interface Engineering

DOI: https://doi.org/10.1038/ncomms15684

This article is licensed under a Creative Commons Attribution 4.0 International License, which permits use, sharing, adaptation, distribution and reproduction in any medium or format, as long as you give appropriate credit to the original author(s) and the source, provide a link to the Creative Commons license, and indicate if changes were made. The images or other third party material in this article are included in the article's Creative Commons license, unless indicated otherwise in a credit line to the material. If material is not included in the article's Creative Commons license and your intended use is not permitted by statutory regulation or exceeds the permitted use, you will need to obtain permission directly from the copyright holder. To view a copy of this license, visit http://creativecommons.org/licenses/by/4.0/

Originators: Grancini, G., Roldán-Carmona, C., Zimmermann, I. *et al.* One-Year stable perovskite solar cells by 2D/3D interface engineering. *Nat Commun* 8, 15684 (2017). https://doi.org/10.1038/ncomms15684

Abstract-Summary

Despite the impressive photovoltaic performances with power conversion efficiency beyond 22%, perovskite solar cells are poorly stable under operation, failing by far the market requirements.

The 2D/3D forms an exceptional gradually-organized multi-dimensional interface that yields up to 12.9% efficiency in a carbon-based architecture, and 14.6% in standard mesoporous solar cells.

This innovative stable and low-cost architecture will enable the timely commercialization of perovskite solar cells.

Introduction

This corresponds, on standard accelerated aging tests, to having <10% drop in PCE for at least 1,000 h. Hybrid perovskite solar cells are still struggling to reach this goal.

Solar cell degradation is not only due by the poor stability of the perovskite layers, but can be also accelerated by the instability of the other layers of the solar cell stack.

This 2D/3D interface brings together the enhanced stability of 2D perovskite with the panchromatic absorption and excellent charge transport of the 3D ones, enabling the fabrication of efficient and ultra-stable solar cells, an important proof of concept for further device optimization and up-scaling.

Within this configuration we demonstrate, for the first time, a remarkable long-term stability of >10,000 h, corresponding to >400 days with zero loss in efficiency over a large-area, fully printable, low-cost and high-efficient solar module of 100 cm^2 (active area of around 50 cm^2) measured under controlled standard conditions and in the presence of oxygen and moisture.

Results

We can rationalize these results suggesting that the 2D/3D perovskite film with 3% AVAI is constituted by a thin layer (possibly a monolayer) of 2D perovskite; an oriented interface where the 3D phase has a marked preferential growth direction, and a pure 3D perovskite arranging in the tetragonal phase on top.

This suggests the presence of a 2D phase mostly retained at the interface with the oxide due to the favourable anchoring of the carboxylic acid group of the AVAI ligand to the TiO_2 scaffold [1].

The emission at 730 nm suggests that a different perovskite phase with a larger band-gap (of 1.69 eV) is formed within the oxide scaffold, but only in the presence of the AVAI precursor.

Our analysis demonstrates the unique role of the 2D perovskite, anchored on the oxide network, in templating the growth of a biphasic 3D $CH_3NH_3PbI_3$: an oriented wider band-gap phase within the oxide scaffold and a standard tetragonal phase on top of it.

Discussion

This leads to an active area of 47.6 cm^2 per module, which is the area considered to calculate the device efficiency.

In our work, the modules consist of 8 cells of 85 $\times$ 7 mm^2, which leads to an active area of 5.95 cm^2 per cell and 47.60 cm^2 per module.

Further optimizations could be done by reducing the interconnect distance between cells, which will probably have two different impacts: less area will be lost, increasing the efficiency per total area in the module and the ohmic losses at the FTO between the interconnect gap will be reduced, improving the efficiency per active area.

The unique combination of the 2D layer acting as a protective window against moisture, preserving the 3D perovskite and of the efficient 3D one provides the hint for the development of a stable perovskite technology, paving the way for the realization of near term high efficient and stable perovskite solar cells for widespread deployment.

Methods

During 30 s. Ten seconds before the end of the program, 100 ml of chlorobenzene were spin-coated on top of the perovskite layer, according to the antisolvent method previously described in literature [2], and finally the perovskite films were sintered at 100 °C during 1 h. Different compositions of 2D–3D perovskite were also tested by mixing different amounts of 1.1 M solution of AVAI:PbI_2 (2:1) with the MAI:PbI_2 precursor solution (0, 3 and 5% molar ratio).

Cells with HTM-free (0.64 cm^2 large) were measured in air, but sealed with the glass slide and Surlyn polymer, the cell temperature during the measurements reaches 60°.

The patterned substrates were coated by a compact TiO_2 layer by aerosol spray pyrolysis, and 1 μm nanoporous TiO_2 or ZrO_2 layer were deposited by screen-printing and sintered at 450 °C for 30 min, as described previously.

Additional information

How to cite this article: Grancini, G. and others One-Year stable perovskite solar cells by 2D/3D interface engineering.

Publisher's note: Springer Nature remains neutral with regard to jurisdictional claims in published maps and institutional affiliations.

Efficient Ambient-Air-Stable Solar Cells with 2D–3D Heterostructured Butylammonium-Caesium-Formamidinium Lead Halide Perovskites

DOI: https://doi.org/10.1038/nenergy.2017.135

Originators: Wang, Z., Lin, Q., Chmiel, F. *et al.* Efficient ambient-air-stable solar cells with 2D–3D heterostructured butylammonium-caesium-formamidinium lead halide perovskites. *Nat Energy* 2, 17135 (2017). https://doi.org/10.1038/nenergy.2017.135

Abstract-Summary

Perovskite solar cells are remarkably efficient; however, they are prone to degradation in water, oxygen and ultraviolet light.

Cation engineering in 3D perovskite absorbers has led to reduced degradation.

2D Ruddlesden–Popper layered perovskites exhibit improved stability, but have not delivered efficient solar cells so far.

We introduce n-butylammonium cations into a mixed-cation lead mixed-halide $FA_{0.83}Cs_{0.17}Pb(I_yBr_{1-y})_3$ 3D perovskite.

Solar cells with an optimal butylammonium content exhibit average stabilized power conversion efficiency of 17.5 ± 1.3% with a 1.61-eV-bandgap perovskite and 15.8 ± 0.8% with a 1.72-eV-bandgap perovskite.

BA/FA/Cs crystallized perovskite thin films

This low value implies that the FA/Cs 3D perovskite phase greatly favours the orientation of the [98] direction parallel to the film normal (equivalently described as a preference for the (h00) planes to be parallel with the sample plane) when crystallizing in the presence of BA.

The reduced lattice constant could be either due to a change in the lattice strain, potentially due to the 3D perovskite grains interfacing with the layered phases, or simply released strain due to the significant reduction in the mosaicity of the film.

We also note that since the halides, FA and Cs cations are relatively diffusive within the perovskite films, it will be feasible for the precise mixed A-site and halide compositions to be dissimilar within the 3D and 2D phases, which should facilitate reduced strain and easier lattice matching.

Photovoltaic performance

To assess whether the BA/FA/Cs perovskite can enhance the solar cell performance, we fabricate devices based on a negative–intrinsic–positive (n–i–p) planar heterojunction architecture.

The champion BA/FA/Cs perovskite device, with a perovskite bandgap of 1.72 eV, exhibits a short-circuit current (J_{sc}) of 19.8 mA cm^{-2}, a fill factor (FF) of 0.73, and an open-circuit voltage (V_{oc}) of 1.18 V, yielding a PCE of 17.2%.

The control device (that is, $x = 0$) exhibits significant J–V hysteresis, which is significantly reduced for the BA/FA/Cs device.

The faster stabilizing rate and the higher SPO ratio is consistent with the perovskite film having a lower density of trap sites responsible for trap-assisted recombination [3], consistent with our measured enhancement in the PL lifetime.

Device stability

To examine whether the BA/FA/Cs perovskite has an impact on device stability we perform ageing tests of high-performance non-encapsulated and encapsulated perovskite devices based on the wide-bandgap (1.72 eV) $FA_{0.83}Cs_{0.17}Pb(I_{0.6}Br_{0.4})_3$ perovskite under full-spectrum simulated sunlight.

We observe that the BA/FA/Cs non-encapsulated devices exhibit enhanced long-term stability, in comparison with the control FA/Cs device in both the J–V PCE and the SPO.

For the BA/FA/Cs perovskite devices, we achieve a champion t_{80} lifetime of 3,880 h for the J–V-measured PCE and 1,680 h for the SPO.

Our results demonstrate a near doubling of the long-term stability in a like-to-like comparison with the control, FA/Cs cells, which were previously the most stable perovskite composition [4–6].

The degradation of the perovskite films is likely to proceed via active defects [7] and hence reducing the number of defects is very likely to positively improve the long-term stability.

Conclusion

We have demonstrated a perovskite composition exhibiting significantly improved device efficiency and operational stability.

With this unique heterostructure, we observe greatly enhanced crystallinity and evidence for reduced defects responsible for non-radiative recombination, which leads to reduced current–voltage hysteresis, enhanced efficiency and significantly improved operational stability in the ensuing solar cells.

Our work represents a significant step towards highly efficient perovskite solar cells having long-term operational stability commensurate with real-world outdoor applications.

Our findings open the possibility of developing much more controlled 2D–3D perovskite heterostructures, and investigating the optoelectronic processes occurring at these interfaces.

Methods

We then added 69 μl of hydroiodic acid (HI; Sigma-Aldrich), 57 wt% in H_2O and 34.6 μl of hydrobromic acid (HBr; Sigma-Aldrich), 48 wt% in H_2O into 1 ml of 1.2 M $FA_{0.83}Cs_{0.17}Pb(I_{0.6}Br_{0.4})_3$ precursor solution, and aged the solution for approximately 48 h under a nitrogen atmosphere without stirring.

For $FA_{0.83}Cs_{0.17}Pb(I_{0.8}Br_{0.2})_3$ perovskite solution, we added 87.2 μl HI acid and 16.5 μl HBr acid to 1 ml of 1.2 M of the precursor solution and was aged for approximately 48 h under a nitrogen atmosphere without stirring.

We spin-coated the doped PCBM solution on top of the as-prepared SnO_2 layer in a nitrogen-filled glove box at 2,500 r.p.m.

On the cleaned FTO substrates, a 2 mg ml^{-1} SnO_2 nanoparticle solution was spin-coated in air at 2,000 r.p.m.

The precursor perovskite solution was spin-coated in a nitrogen-filled glove box at 2,000 r.p.m.

Additional Information

How to cite this article: Wang, Z. et al. Efficient ambient-air-stable solar cells with 2D–3D heterostructured butylammonium-caesium-formamidinium lead halide perovskites.

Publisher's note: Springer Nature remains neutral with regard to jurisdictional claims in published maps and institutional affiliations.

[Section 4.7]

The archetypal MA-based perovskite solar cells (that is, methylammonium lead triiodide: $MAPbI_3$) exhibit intrinsic thermal instability due to the relatively volatile MA cations, which are released from the film during heating, and this process is accelerated in the presence of moisture [8–10].

Until recently, perovskite solar cells employing layered perovskite compounds have exhibited relatively low efficiency (PCE < 10%) (refs [11, 12]), assumed to be due to the inhibition of out-of-plane charge transport by the insulating spacer cations.

A recent study adopted a 'hot-casting' technique to fabricate highly crystalline $(BA)_2(MA)_3Pb_4I_{13}$ (BA: n-butylammonium, $C_4H_9NH_3$) Ruddlesden–Popper phase perovskite films with the crystalline planes aligned along the out-of-plane orientation, which facilitates efficient charge transport along the perovskite planes, achieving a 12.5%-efficient layered perovskite solar cell with improved moisture and light stability [13].

Our work illustrates that engineering heterostructures between 2D and 3D perovskite phases is possible, and can lead to enhancement of both performance and stability of perovskite solar cells.

Origin of Vertical Orientation in Two-Dimensional Metal Halide Perovskites and its Effect on Photovoltaic Performance

DOI: https://doi.org/10.1038/s41467-018-03757-0

This article is licensed under a Creative Commons Attribution 4.0 International License, which permits use, sharing, adaptation, distribution and reproduction in any medium or format, as long as you give appropriate credit to the original author(s) and the source, provide a link to the Creative Commons license, and indicate if changes were made. The images or other third party material in this article are included in the article's Creative Commons license, unless indicated otherwise in a credit line to the material. If material is not included in the article's Creative Commons license and your intended use is not permitted by statutory regulation or exceeds the permitted use, you will need to obtain permission directly from the copyright holder. To view a copy of this license, visit http://creativecommons.org/licenses/by/4.0/.

Originators: Chen, A.Z., Shiu, M., Ma, J.H. *et al.* Origin of vertical orientation in two-dimensional metal halide perovskites and its effect on photovoltaic performance. *Nat Commun* 9, 1336 (2018). https://doi.org/10.1038/s41467-018-03757-0

Abstract-Summary

Simple solution processing of the 2D perovskite provides opportunities for manufacturing devices at drastically lower cost compared to current commercial technologies.

A key to high device performance is to align the 2D perovskite layers, during the solution processing, vertical to the electrodes to achieve efficient charge transport.

It is yet to be understood how the counter-intuitive vertical orientations of 2D perovskite layers on substrates can be obtained.

We also demonstrate control over the degree of preferential orientation of the 2D perovskite layers and its drastic impact on device performance.

Introduction

For devices such as solar cells and LEDs, wherein the MHP thin films are sandwiched between electrodes, vertical alignment of 2D MHPs is desired to obtain efficient charge transport to the electrodes [13, 14].

The self-assembly processes through which 2D MHPs can be aligned vertically remain a mystery and cannot yet be controlled.

We reveal a mechanism responsible for the formation of vertically orientated 2D MHPs in thin films using grazing incidence wide angle X-ray scattering (GIWAXS).

The formation of vertically orientated 2D MHPs at the anisotropic environment of the liquid–air interface is likely due to preference for the aliphatic chain of butylammonium molecules bound to the metal halide slabs to remain in the solution environment.

Knowing the origin of the preferential orientation allows us to rationally tune the degree of vertical orientation of 2D MHP with similar thin film morphology.

Results

The observed strong vertical orientation in our films suggest that homogeneous nucleation is suppressed and, instead, heterogeneous nucleation at an interface is dominant.

The lack of dependence on different substrates as well as the formation of strong vertical orientation regardless of the tortuous and uneven surface of the mesoporous TiO_2 substrate indicates that the nucleation does not occur at the substrate–liquid interface.

The confirmation of the crystallization at the liquid–air interface sheds light on the origin of the peculiar preferential vertical orientations.

The formation of vertically oriented $BA_2MA_3Pb_4I_{13}$ also indicates that a monolayer of butylammonium molecules as typically seen in Langmuir–Blodgett films [15] with butyl chains vertical to the liquid–air interface, likely does not form in this situation.

The pre-crystallization annealed film with a strong vertical orientation outperforms the films with partial random orientation due to a significantly higher short circuit current density (J_{sc}) and lower series resistance (R_s), while both devices have similar open circuit voltage (V_{oc}).

Discussion

We reveal that vertically oriented 2D MHP $BA_2MA_3Pb_4I_{13}$ nucleates and grows from the precursor liquid–air interface.

By comparing electronic properties of vertically oriented and partially randomly oriented thin films, we find that a strong vertical orientation is necessary for optimal solar cell performance.

This insight on the 2D MHP self-assembly and its impact on optoelectronic device performance will accelerate the progress toward a wide-spread application of 2D MHPs devices.

Methods

The solution was filtered with a 0.22 μm filter and then deposited on the substrate by spinning at 2000 r.p.m.

After about 20 s of spin-coating the slide was taken off the spincoater and immediately placed on a hotplate to anneal at 120 °C for 10 min, where the precursor crystallized and turned dark brown.

Post-crystallization annealing method: 0.75 M $BA_2MA_3Pb_4I_{13}$ solution dissolved in DMAc was filtered with a 0.22 μm filter and then deposited on the substrate by spinning at 2000 r.p.m.

After 60 s of spin-coating the slide had already dried and turned dark brown, and was taken off the spincoater and placed on a hotplate to anneal at 120 °C for 10 min.

A layer of PCBM was deposited by spin-coating a freshly filtered 10 mg ml^{-1} solution in chloroform at 3000 r.p.m.

After the slides cooled down, a layer of BCP was spin-coated at 4000 r.p.m.

2D Perovskite Stabilized Phase-Pure Formamidinium Perovskite Solar Cells

DOI: https://doi.org/10.1038/s41467-018-05454-4

This article is licensed under a Creative Commons Attribution 4.0 International License, which permits use, sharing, adaptation, distribution and reproduction in any medium or format, as long as you give appropriate credit to the original author(s) and the source, provide a link to the Creative Commons license, and indicate if changes were made. The images or other third party material in this article are included in the article's Creative Commons license, unless indicated otherwise in a credit line to the material. If material is not included in the article's Creative Commons license and your intended use is not permitted by statutory regulation or exceeds the permitted use, you will need to obtain permission directly from the copyright holder. To view a copy of this license, visit http://creativecommons.org/licenses/by/4.0/.

Originators: Lee, JW., Dai, Z., Han, TH. *et al.* 2D perovskite stabilized phase-pure formamidinium perovskite solar cells. *Nat Commun* 9, 3021 (2018). https://doi.org/10.1038/s41467-018-05454-4

Abstract-Summary

Compositional engineering has been used to overcome difficulties in fabricating high-quality phase-pure formamidinium perovskite films together with its ambient instability.

We report the fabrication of phase-pure formamidinium-lead tri-iodide perovskite films with excellent optoelectronic quality and stability.

Incorporation of 1.67 mol% of 2D phenylethylammonium lead iodide into the precursor solution enables the formation of phase-pure formamidinium perovskite with an order of magnitude enhanced photoluminescence lifetime.

Introduction

Compositional engineering has been considered an important approach to enhance the stability and performance of PVSK solar cells.

Incorporation of smaller inorganic 'A' cations, such as Cs and Rb, has further improved the stability and PCE of the PVSK solar cells with the lowest open-circuit voltage (V_{OC}) deficit of 0.39 V [16, 17].

Their performance and stability are still relatively poor comparing with those of $MAPbI_3$ or mixed-cation-halide PVSK solar cells.

Thanks to the superior optoelectronic quality, we were able to fabricate a PVSK solar cell with a stabilized efficiency of 20.64% (certified stabilized efficiency of 19.77%).

The PVSK solar cell shows a peak V_{OC} of 1.130 V, corresponding to a loss-in-potential of 0.35 V considering the E_g of 1.48 eV versus 0.39 V for mixed-cation-halide perovskite solar cells [17].

Results

With addition of 1.67 mol% 2D PVSK, proportion of the fast decay component (τ_1) was decreased (from 51.8% to 46.5%) while τ_2 significantly elongated from 78.5 ns to 148.7 ns, which indicates reduced defect density and enhanced charge carrier lifetime.

The enhanced device performance with 2D PVSK is mainly due to improved FF and V_{OC}, which can be attributed to improved phase purity and elongated carrier lifetime with reduced defect density, facilitating carrier transport and reducing the charge recombination [16].

After 500 h of exposure, the control device degraded to ca 52.3% of its initial PCE whereas the target device maintained 72.3% of the initial PCE, indicating enhanced stability with addition of 2D PVSK.

This indicates the stability of the device was significantly improved with addition of 2D PVSK.

Discussion

We demonstrated a reproducible way to fabricate phase-pure formamidinium triiodide PVSK with high-optoelectronic quality and stability by incorporating 2D PVSK.

The large phenylethylammonium molecules from 2D PVSK precursors interact with $FAPbI_3$ crystals to facilitate formation of the cubic PVSK phase during crystallization, which subsequently functionalize the grain boundaries after completion of the crystallization.

Owed to the functionalized grain boundaries by the 2D PVSK, the phase stability of the film under high RH significantly improved and migration of ions (or charged defects) was suppressed, resulting in significantly improved ambient and operational stability of the device.

We believe our approach to utilize spontaneously formed grain boundary 2D PVSK will provide important insights for the research community to design PVSK materials to achieve record PCEs accompanied by high stability and longevity.

Methods

The PVSK solution was spin-coated at 4000 rpm for 20 s where 0.15 mL of diethyl ether (anhydrous, >99.0%, contains BHT as stabilizer, Sigma-Aldrich) was dropped after 10 s on the spinning substrate.

The resulting transparent adduct film was heat-treated at 100 °C for 1 min followed by 150 °C for 10 min. (for the best performing target device, the annealing condition was adjusted to 80 °C 1 min followed by 150 °C for 20 min) The spiro-MeOTAD solution was prepared by dissolving 85.8 mg of spiro-MeOTAD (Lumtec) in 1 mL of chlorobenzene (anhydrous, 99.8%, Sigma-Aldrich) which was doped by 33.8 μl of 4-tert-butylpyridine (96%, Aldrich) and 19.3 μl of Li-TFSI (99.95%, Aldrich, 520 mg mL^{-1} in acetonitrile) solution.

For light exposure under open-circuit condition, the encapsulated devices were exposed to ca 0.9 sun (90 ± 5 mW cm^{-2}) generated by halogen lamps under open-circuit condition, of which steady-state PCE was periodically measured with different exposure time under 1 sun illumination.

Dimensionality Engineering of Hybrid Halide Perovskite Light Absorbers

DOI: https://doi.org/10.1038/s41467-018-07382-9

This article is licensed under a Creative Commons Attribution 4.0 International License, which permits use, sharing, adaptation, distribution and reproduction in any medium or format, as long as you give appropriate credit to the original author(s) and the source, provide a link to the Creative Commons license, and indicate if changes were made. The images or other third party material in this article are included in the article's Creative Commons license, unless indicated otherwise in a credit line to the material. If material is not included in the article's Creative Commons license and your intended use is not permitted by statutory regulation or exceeds the permitted use, you will need to obtain permission directly from the copyright holder. To view a copy of this license, visit http://creativecommons.org/licenses/by/4.0/.

Originators: Gao, P., Bin Mohd Yusoff, A.R. & Nazeeruddin, M.K. Dimensionality engineering of hybrid halide perovskite light absorbers. *Nat Commun* 9, 5028 (2018). https://doi.org/10.1038/s41467-018-07382-9

Abstract-Summary

Hybrid halide perovskite solar cells were first demonstrated in 2009 with cell efficiency quickly soaring from below 10% to more than 23% in a few years.

Halide perovskites have the desirable processing simplicity but are very fragile when exposed to water and heat.

Introduction

The last two strategies to stabilize perovskite can be generally called multi-dimensional perovskite (MDP).

In most instances, the gain in the long-term stability was accompanied by the sacrifice in the power conversion efficiency (PCE) except the strategies represented by mixing three-dimensional (3D) perovskites with low-dimensional (2D, 1D, or 0D) perovskites (LDP) [18–20] to form MDP with a general formula of $(A)_2(CH_3NH_3)_{n-1}M_nX_{3n+1}$ ($1 \leq n \leq \infty$).

As one of the three main motivations [21] behind the development of derivatives of hybrid halide perovskites, the MDP strategy showed the potential to provide parallelly improved stability and PCEs.

Comparing with other approaches to stabilize 3D perovskite materials (vide supra), the MDP approach has the advantages of convenient, less cost, and taking effect in situ from inside of the materials.

The evolution of dimensionality in hybrid halide perovskite

Stoumpos and othersbutyl-ammonium [20], managed to solve the crystal structures of $(BA)_2(MA)_{n-1}Pb_nI_{3n+1}$ (BA: butyl-ammonium; MA: methyl ammonium; n equals 1 to 5).

The visualized structure is beneficial to understand the stage of quasi-2D layered perovskite during the transition from monolayer 2D perovskite to infinite 3D perovskite after tuning the organic cation and stoichiometry of precursors.

In the case of $(BA)_2(MA)_{n-1}Pb_nI_{3n+1}$ series, the sharp optical absorption edges increase in energy with decreasing n value from 1.50 eV (n equals ∞) to 2.43 eV (n equals 1) in the same manner of the dimension reduction [22–24].

Yao and others [12], prepared 'quasi 2D' perovskite compounds $(PEI)_2(MA)_{n-1}Pb_nI_{3n+1}$ by incorporating polyethyleneimine (PEI) cations within the layered structure.

In this work when n equals 3, the resulted E_g is slightly smaller than that of $(BA)_2(MA)_2Pb_3I_{10}$ due to stronger interactions between the organic−inorganic structures when multi ammonium cation is used [25, 26].

Interplay between dimensionality, PCE, and stability

In Class I MDP, the bulky organic cations (A) fragment the compact 3D perovskite structure into isolated inorganic layers of corner-sharing $[MX_6]^{4-}$ octahedrons with certain layer number (n) [27].

A thin $(BA)_2PbI_4$ layer formed on the surface and at the grain boundaries of 3D perovskite can enhance the thermal stability of the devices and increase the PCE to 19.56% [28].

While synchronously increased stability and PCE were observed in the Class I MDP based devices, there is an apparent disparity between their PCE and that of 3D perovskites.

In the region of Class I MDP, PCEs usually are compromised for the stability with the highest reported PCE of 15.4%, contrasting to the state-of-the-art PCE from 3D perovskite of 23.3% [29, 30].

Like the previous Class II MDP, with this innovation, they realized an excitingly high PCE of 19.89% in a p-i-n device and much-increased moisture/thermal stability as well as suppressed cross-layer ion migration.

Stability enhancement with MDP

According to the differences (0.36 eV) between the energy required to remove PEAI from the perovskite and that for MAI, they conclude that the adoption of MDP (either Class I or Class II) over 3D Perovskite can reduce the desorption rate by six orders of magnitude, and slowdown film decomposition by 1000-fold.

Without considering the testing conditions like humidity and temperature, the data shows that although 3D perovskites were reported with unexpectedly good stability under test, MDP perovskites were tested with the twice longer time span.

Although the ever-extending span of lifetime test is exciting, one must be soberly aware that most of the stability measurements were not carried out under the same standard conditions, e.g., unified relative humidity, temperature, light intensity, atmosphere and the same fitting function of PCE retention.

Given the facts above, it is necessary to consolidate the testing conditions for the evaluation of the stability of a perovskite solar cell.

Conclusions and perspectives

Comparing to pure 3D and LD perovskites, multi-dimensional perovskites (MDP) maintain balanced performances, bandgaps, charge transporting properties, and ambient stability, which are critical for photovoltaic applications.

Techniques like hot-casting or so-called 'pre-crystallization annealing' should be adopted to induce vertical growth of perovskite crystallites since the intrinsic 2D perovskites tend to grow parallel to the plane of the substrate and the charge transfer efficiency in the normal direction (to the charge selective contact layers) is very low.

The Class I MDPs with Ruddlesden–Popper perovskite structure have provided devices with improved stability and moderate PCEs, whereas interfacial modifications and passivation in Class II MDPs further push the device PCE up to 20% and on par with 3D perovskite [31].

Sharing the merits of the low-cost manufacturing property of 3D perovskites, such as roll-to-roll slot die-coating, screen-printing, or ink-jet printing technologies, large area MDP modules with improved ambient stability will be the real breakthrough and challenger to the state-of-the-art Si-based PVs.

Tailoring Vertical Phase Distribution of Quasi-Two-Dimensional Perovskite Films via Surface Modification of Hole-Transporting Layer

DOI: https://doi.org/10.1038/s41467-019-08843-5

This article is licensed under a Creative Commons Attribution 4.0 International License, which permits use, sharing, adaptation, distribution and reproduction in any medium or format, as long as you give appropriate credit to the original author(s) and the source, provide a link to the Creative Commons license, and indicate if changes were made. The images or other third party material in this article are included in the article's Creative Commons license, unless indicated otherwise in a credit line to the material. If material is not included in the article's Creative Commons license and your intended use is not permitted by statutory regulation or exceeds the permitted use, you will need to obtain permission directly from the copyright holder. To view a copy of this license, visit http://creativecommons.org/licenses/by/4.0/.

Originators: Liu, T., Jiang, Y., Qin, M. *et al.* Tailoring vertical phase distribution of quasi-two-dimensional perovskite films via surface modification of hole-transporting layer. *Nat Commun* 10, 878 (2019). https://doi.org/10.1038/s41467-019-08843-5

Abstract-Summary

Vertical phase distribution plays an important role in the quasi-two-dimensional perovskite solar cells.

We report that the vertical distribution of layer numbers in the quasi-two-dimensional perovskite films deposited on a hole-transporting layer is different from that on glass substrate.

The vertical distribution could be explained by the sedimentation equilibrium because of the colloidal feature of the perovskite precursors.

A self-assembly layer is used to modify the acidic surface property of the hole-transporting layer that induces the appearance of desired vertical distribution for charge transport.

Introduction

The layered quasi-two-dimensional (Q-2D) $(A_1)_2(A_2)_{n-1}B_nX_{3n+1}$ [where A_1 is large-size bulky organic cation, such as butylammonium (BA) or phenethylammonium (PEA); A_2 is Cs, methylammonium (MA) or formamidinium (FA); B is Pb or Sn; X is halide anion; and the integer n is number of layered inorganic [PbI_6] sheets between two bulky organic cation spacers] perovskites have shown potential to improve the physico-chemical stability [31, 33–35].

We find that the desired vertical phase distribution of n from the small to the large disappears when the Q-2D perovskite is deposited on the hole-transporting layer (HTL) of poly(3,4-ethylenedioxythiophene): poly(styrene sulfonate) (PEDOT:PSS), which is different from that on the glass substrate.

We employ a self-assembly layer of 4-bromobenzenediazonium tetrafluoroborate covalently anchored onto the surface of the PEDOT:PSS that can induce the appearance of the desired vertical phase distribution of n in the Q-2D perovskite films.

Results

We tried to build this vertical phase distribution on PEDOT:PSS films that are typically used as HTL in the inverted Q-2D perovskite solar cells.

The PEDOT:PSS suppresses the vertical phase distribution of the Q-2D perovskite.

When deposited on the PEDOT:PSS, the vertical distribution of n from the small (bottom side) to the large (top side) disappears in the Q-2D perovskite films.

We deposited Q-2D perovskite films on PEDOT:PSS, PSSH and DMSO treated PEDOT:PSS and measured their PL from the back side.

The vertical distribution of n can be obtained when the Q-2D perovskite is deposited on the BrB-PEDOT:PSS.

The Q-2D perovskite films on BrB-PEDOT:PSS show the same set of diffraction spots as that on PEDOT:PSS substrate, but stronger intensity.

This also confirms the important role of the BrB-PEDOT:PSS tailoring the vertical phase distribution in Q-2D perovskite films.

Discussion

We have reported that the vertical phase distribution of Q-2D perovskite films on glass substrates can be explained by the sedimentation equilibrium because of the colloidal feature of the precursors, and can be tailored by the surface modification and acid addition.

Main observations and conclusions include: (1) the vertical phase distribution of Q-2D perovskite films on PEDOT:PSS HTL is different from the glass substrates.

More small-n phases in the Q-2D perovskite films form at the bottom side (close to substrate) when deposited on glass substrates, compared to the Q-2D perovskite films deposited on PEDOT:PSS layers; (2) The origin of the vertical phase distribution of n is the colloid property of the perovskite precursors (the colloid size on the order of 100–1000 nm) that follows the sedimentation equilibrium.

The vertical distribution with different n can hardly be observed in the Q-2D perovskite films when deposited on PEDOT:PSS.

Methods

For optical (absorption and PL) and morphology (SEM, GIWAXS) characterization, the perovskite samples were deposited on different substrates (glass, PEDOT:PSS, BrB-PEDOT:PSS) via the hot-casting method that is the same as that used for the devices.

The PL measurements were performed using a fluorescence spectrometer (Edinburgh Instruments FLS920) with a 150 W Xe lamp as an excitation source at 450 nm.

In-depth XPS (AXIS-ULTRA DLD-600W) profiling was performed on the samples of glass/HTL/Q-2D, where the HTL is PEDOT:PSS or BrB-PEDOT:PSS.

Synthetic Control over Orientational Degeneracy of Spacer Cations Enhances Solar Cell Efficiency in Two-Dimensional Perovskites

DOI: https://doi.org/10.1038/s41467-019-08980-x

This article is licensed under a Creative Commons Attribution 4.0 International License, which permits use, sharing, adaptation, distribution and reproduction in any medium or format, as long as you give appropriate credit to the original author(s) and the source, provide a link to the Creative Commons license, and indicate if changes were made. The images or other third party material in this article are included in the article's Creative Commons license, unless indicated otherwise in a credit line to the material. If material is not included in the article's Creative Commons license and your intended use is not permitted by statutory regulation or exceeds the permitted use, you will need to obtain permission directly from the copyright holder. To view a copy of this license, visit http://creativecommons.org/licenses/by/4.0/.

Originators: Hu, J., Oswald, I.W.H., Stuard, S.J. *et al.* Synthetic control over orientational degeneracy of spacer cations enhances solar cell efficiency in two-dimensional perovskites. *Nat Commun* 10, 1276 (2019). https://doi.org/10.1038/s41467-019-08980-x

Abstract-Summary

How the chemical nature of the organic cations affects the properties of two-dimensional perovskites and devices is rarely reported.

We demonstrate that the selection of spacer cations (i.e., selective fluorination of phenethylammonium) affects the film properties of two-dimensional perovskites, leading to different device performance of two-dimensional perovskite solar cells (average $n = 4$).

This work provides key missing information on how spacer cations exert influence on desirable electronic properties and device performance of two-dimensional perovskites via the weak and cooperative interactions of these cations in the crystal lattice.

Introduction

2D OIHP solar cells have recently been demonstrated with significant high efficiency via a couple of methods such as hot-casting [13] and using additives [36, 37, 43–45], due to presumably achieved vertical alignment of the inorganic slabs.

Due to the easy accessibility of simple aliphatic ammoniums (e.g., butylammonium, C_4H_9-NH_3^+, BA), most 2D OIHP based solar cells of high efficiency have employed them as spacer cations [13, 43, 44].

While we observe over 10% photovoltaic efficiency when 3-fluorophenethylammonium (mF1PEA) or 4-fluorophenethylammonium (pF1PEA) is used as the organic cation in 2D OIHP based solar cells, the efficiency of solar cells based on 2-fluorophenethylammonium (oF1PEA) is less than 1%.

To further understand how the organic cation would affect the structure of these 2D OIHPs, we analyze single crystals of 2D OIHPs ($n = 1$) with these fluorinated PEA cations and disclose that all crystals have a similar inorganic framework structure, yet very different organic cation packing arrangements.

Results

These observations indicate in PEA, mF1PEA, or pF1PEA based 2D OIHP films, a vertical phase distribution of multiple 2D OIHP phases occurs, where phases having smaller n values dominate on the glass side of the substrate while phases with larger n or 3D appear more on the air side.

We propose that during the film formation of the oF1PEA based 2D OIHP film, the lack of ordering of the organic bilayer would cause the formation of small crystal domains and may also play a part in why the film morphology and subsequent device performance was so drastically lower than either PEA, pF1PEA, or mF1PEA 2D OIHP films.

The small crystal domain and random orientation in oF1PEA 2D OIHP films, caused by the unfavorable formation energy and highly disordered crystal structure, would further hinder the device performance.

Discussion

While three OIHP films (PEA, mF1PEA and pF1PEA) show more or less a vertical phase distribution where the small n phases accumulated near the substrate and 3D phases near the surface of the film, the oF1PEA OIHP film only shows a rather random distribution of 2D phases.

These differences could be caused by the organic cation interactions (e.g., dipole–dipole interactions), which is likely the fundamental reason for the observed photovoltaic device differences among these different PEA cations based 2D OIHP films.

We note cations with single benzene ring like PEA lack the coveted optoelectronic functionalities; however, this work still provides valuable insight to the future design of more complex conjugated cations to achieve desirable properties for a variety of applications including photovoltaics [11, 36, 37], light emitting [34, 38, 39], transistors [40, 41], spin electronics [42], etc We believe future design of organic cations and understanding the intermolecular interactions associated with these organics, will play an essential role in further development of these 2D OIHP materials.

Methods

For 2D OIHP films, the precursor solution was made by dissolving PEAI (or oF1PEAI, mF1PEAI, pF1PEAI), MAI and PbI_2 (99.9985%, from Alfa Aesar) in dimethylformamide (DMF) (from Sigma-Aldrich) with the molar ratio of PEAI:MAI:PbI_2 = 2:3:4 and the mixture was stirred for at least 60 min.

The 2D OIHP film was spin-coated at 5000 rpm for 20 s from the precursor solution on a substrate (glass substrates for XRD and optical characterizations; poly(3,4-ethylenedioxythiophene) polystyrene sulfonate (PEDOT:PSS) coated indium doped tin oxide (ITO) substrates for solar cells, SEM and AFM; Si wafer for ToF-SIMS; PEDOT:PSS coated Si wafer for GIWAXS) and then the resulting film was quickly transferred to a hot plate at desired temperature (65 °C for pF1PEA and 40 °C for the rest) to anneal 30 s. The spin-coating process was done in an ambient condition with a relative humidity around 10–45%.

Efficient and Stable Ruddlesden–Popper Perovskite Solar Cell with Tailored Interlayer Molecular Interaction

DOI: https://doi.org/10.1038/s41566-019-0572-6

Originators: Ren, H., Yu, S., Chao, L. *et al.* Efficient and stable Ruddlesden–Popper perovskite solar cell with tailored interlayer molecular interaction. *Nat. Photonics* 14, 154–163 (2020). https://doi.org/10.1038/s41566-019-0572-6

Abstract-Summary

Two-dimensional Ruddlesden–Popper phase (2DRP) perovskites are known to exhibit improved photostability and environmental stability compared with their three-dimensional (3D) counterparts.

Fundamental questions remain over the interaction between the bulky alkylammoniums and the 2DRP perovskite framework.

The result is 2DRP perovskite solar cells with significantly improved efficiency and stability.

Main

Compared with 3D perovskites, 2DRP perovskites have a unique degree of freedom that can be exploited to improve PSC efficiency—the bulky ammonium RNH_3 layer.

This interaction controls the assembly of 2DRP layered perovskite segments and the charge transport across the segments [46].

Investigations of this interaction are expected to enable enhancement of stability and the optoelectronic conversion efficiency of 2DRP perovskite films.

We report a joint experimental–theoretical study on tailoring the interlayer molecular interaction among 2DRP layered perovskite segments and investigating its impact on the efficiency and stability of 2DRP PSCs.

The S–S interaction-induced interlayer interaction enabled smooth, dense and pinhole-free 2DRP perovskite films with a low trap state density and allowed an efficient charge transfer process.

The strengthened interlayer molecular interaction significantly improved the environmental and thermal stability of the 2DRP perovskite films.

Results and discussion

We found a probable S–S interaction in the 2DRP $(MTEA)_2(MA)_4Pb_5I_{16}$ (n = 5) perovskites via X-ray photoelectron spectroscopy (XPS).

The results strongly demonstrate that the S–S interaction in MTEA can stabilize and tighten the framework of the 2DRP perovskite phase to allow vertically oriented growth.

To explore whether the S–S interaction can enhance the photoelectronic properties of the $(MTEA)_2(MA)_4Pb_5I_{16}$ over $(BA)_2(MA)_4Pb_5I_{16}$ (n = 5) 2DRP perovskites, the trap state density, charge carrier mobility and charge transfer kinetics were investigated.

The faster photoluminescence (PL) decay in the PEDOT:PSS and $PC_{61}BM$ heterojunctions with 2DRP perovskites further indicates the effective charge extraction of $(MTEA)_2(MA)_4Pb_5I_{16}$ compared to $(BA)_2(MA)_4Pb_5I_{16}$, which is related to the highly oriented RP phases in the $(MTEA)_2(MA)_4Pb_5I_{16}$ 2DRP perovskites.

The flat, dense and highly oriented crystalline 2DRP perovskite thin films of $(MTEA)_2(MA)_4Pb_5I_{16}$, with their excellent optical and electrical properties, are particularly suitable for preparing photovoltaic devices.

Conclusions

We have shown that interactions among the organic tails of the ammonium molecules play an important role in constructing efficient and stable 2DRP hybrid perovskites.

Efficient intercalated 2DRP hybrid perovskites stabilized by the S–S interaction are reported.

The discovered interlayer interactions of 2DRP perovskites provide a particularly flexible framework and an alternative opportunity to tailor this scientifically and technologically interesting class of hybrid perovskites for potential applications.

Methods

For 50 s and annealed at 120 °C for 30 min. $(MTEA)_2(MA)_4Pb_5I_{16}$ and $(BA)_2(MA)_4Pb_5I_{16}$ films were fabricated by spin-coating at 4,000 r.p.m.

The electron transport layer $PC_{61}BM$ in chlorobenzene solution (18 mg ml^{-1}) was deposited onto the perovskite layer by spin-coating at 1,000 r.p.m.

For the thermal stability test, the two types of 2DRP perovskite film were placed on a hotplate at 85 °C or 150 °C in a N_2-filled glovebox.

For the light soaking stability tests, the light intensity and device temperature were fixed at 1 sun, AM1.5G illumination and 25 °C, respectively.

The stability of the device over time was determined at the maximum power point under simulated AM1.5G light (Enlitech SS-F5-3A) in a N_2-filled glovebox for 1,000 h. The UV–vis absorbance spectra of all samples were measured using an Agilent Cary 5000 UV–vis spectrophotometer with an integrating sphere and a Hitachi F-4600 fluorescent spectrophotometer.

Reduced-Dimensional Perovskite Photovoltaics with Homogeneous Energy Landscape

DOI: https://doi.org/10.1038/s41467-020-15451-1

This article is licensed under a Creative Commons Attribution 4.0 International License, which permits use, sharing, adaptation, distribution and reproduction in any medium or format, as long as you give appropriate credit to the original author(s) and the source, provide a link to the Creative Commons license, and indicate if changes were made. The images or other third party material in this article are included in the article's Creative Commons license, unless indicated otherwise in a credit line to the material. If material is not included in the article's Creative Commons license and your intended use is not permitted by statutory regulation or exceeds the permitted use, you will need to obtain permission directly from the copyright holder. To view a copy of this license, visit http://creativecommons.org/licenses/by/4.0/.

Originators: He, T., Li, S., Jiang, Y. *et al.* Reduced-dimensional perovskite photovoltaics with homogeneous energy landscape. *Nat Commun* 11, 1672 (2020). https://doi.org/10.1038/s41467-020-15451-1

Abstract-Summary

Reduced-dimensional (quasi-2D) perovskite materials are widely applied for perovskite photovoltaics due to their remarkable environmental stability.

We thus propose to form homogeneous energy landscape to overcome the problem.

The lowest energy distribution and homogeneous energy landscape are achieved through carefully regulating their crystallization kinetics.

We conclude that homogeneous energy landscape significantly reduces the Shockley-Read-Hall recombination and suppresses the quasi-Fermi level splitting, which is crucial to achieve high V_{oc}.

Introduction

Energy transfer dynamics is demonstrated to dominate the carrier transport in quasi-2D films, because the cascade energy transfer typically happens at a very early stage after photoexcitation.

Severe energy disorder exists in the quasi-2D perovskite films, which is mainly induced by the inhomogeneous energy landscape [47].

It is foreseen that quasi-2D perovskite photovoltaics should suffer from the energy disorder as well.

Quasi-2D perovskite film with homogeneous energy landscape is proposed here to address the problems.

Achieving homogeneous energy landscape in low value regions is necessary to ensure both performance and stability of the PSCs.

Narrow value-distributed quasi-2D films are formed, once the organic cations become homogeneously distributed.

Electrical characterizations confirm that the homogeneous energy landscape decreases the defect density and reduces the film's energy disorder.

Results

The quasi-2D perovskite films' formation kinetics need to be well controlled to achieve narrow value distribution.

Debye-Scherrer rings, (1 1 0) and (2 2 0) planes, belonging to high values or 3D perovskites, were observed at the bottom of BA and pristine MAMP films with a larger grazing-incidence angle.

It is worth mentioning that ΔE represents the narrowest energy distribution for low value quasi-2D perovskite films reported to date.

Graded MAMP quasi-2D perovskite films with homogeneous energy landscape have been achieved here.

APSA-treated MAMP film exhibited one order of magnitude higher electron mobility, 2.13×10^{-2} cm^2 V^{-1} s^{-1}, indicating more efficient minority carrier transport realized in this homogeneous energy landscape.

Low value species tended to stay on the top in APSA-treated MAMP film.

Discussion

We have demonstrated that the homogeneous energy landscape is essential to achieve high-performance quasi-2D perovskite photovoltaics, especially high V_{oc}.

A homogeneous energy landscape is formed in APSA-treated MAMP film.

We correlate the relationship between the energy landscape homogeneity and device performance, and further investigate the physical origin of low V_{loss}.

Owing to the reduced nonradiative recombination induced by the homogeneous energy landscape, the device achieves an excellent PCE of 16.53%, accompanied by a high V_{oc} of 1.21 V. In addition, the device exhibits significantly improved stability compared with traditional hot-casting films, due to the more ideal vertical phase alignment and absent high value phases.

Methods

For $MAMPMA_{n-1}Pb_nI_{3n+1}$ SCs, different ratios of PbO, MAI, and MAMPI mixture were dissolved in 30 mL of HI solution at about 110 °C.

The precursor solutions were spin-coated on top of the TiO_2 film at 4000 r.p.m.

GIWAXS patterns of quasi-2D perovskite films were measured on the beamline 1W2A at Beijing Synchrotron Radiation Facility (BSRF), China.

All films were measured via the same exposure time of 240 s. GIWAXS patterns were collected at a beam energy of 10 keV by an area detector with 3450 × 3450-pixel resolution.

TA measurements of quasi-2D films were carried out by a Helios pump–probe system (Ultrafast Systems LLC) with an amplified femtosecond laser system (Coherent, 35 fs, 7 mJ per pulse, 1 kHz, 800 nm).

Focusing the 800-nm beams (split from the regenerative amplifier with a tiny portion, 400 nJ per pulse) onto a sapphire plate produced the white-light continuum (WLC) probe pulses.

Two-Dimensional Ruddlesden–Popper Layered Perovskite Solar Cells Based on Phase-Pure Thin Films

DOI: https://doi.org/10.1038/s41560-020-00721-5

Originators: Liang, C., Gu, H., Xia, Y. *et al.* Two-dimensional Ruddlesden–Popper layered perovskite solar cells based on phase-pure thin films. *Nat Energy* 6, 38–45 (2021). https://doi.org/10.1038/s41560-020-00721-5

Abstract-Summary

Such perovskites typically consist of multiple quantum wells with a random well width distribution.

We report phase-pure quantum wells with a single well width by introducing molten salt spacer n-butylamine acetate, instead of the traditional halide spacer n-butylamine iodide.

Due to the strong ionic coordination between n-butylamine acetate and the perovskite framework, a gel of a uniformly distributed intermediate phase can be formed.

Main

Two-dimensional Ruddlesden–Popper layered metal-halide perovskites, with desirable optoelectronic properties and good environmental stability, have recently attracted extensive research interest as one of the most promising materials for the fabrication of long-term operationally stable and high-performance perovskite solar cells (PSCs) [39–44, 47].

Layered perovskites deposited using solution methods typically suffer from a mixture of multiple quantum wells with a random well width (n number) distribution (MQWs), which has severely hindered further progress in this field [13–49].

The perovskite solution prepared with stoichiometric amounts of precursors, however, has been confirmed to be a complex mixture of colloid particles with a random size distribution in a mother solution of dissolved ions [50, 51].

The deposition of phase-pure QWs remains a key scientific challenge for the thermodynamic stability of compounds in solution.

For strong ionic coordination between BAAc and the perovskite framework during the initial stage of solution processing, compounds with a uniformly distributed intermediate phase can be gelled from the precursor solution with near-monodisperse unit cell particles.

Properties of MQW and phase-pure QW films

These results reveal that near-single-phase perovskite QWs films could be deposited with a molten salt spacer.

This increase combined with an efficient charge carrier funnelling from the thin to thicker QWs [34, 39, 48, 52–56] causes the PL of the MQW films with > 2 to be dominated by the broad emission from 3D perovskite, with a slight contribution from thin QWs.

These results further confirm that near-single-phase perovskite QW films were deposited with the molten salt spacer BAAc.

The 3D ($n = \infty$) phase on top of the films is subject to degradation, which is not good for perovskite stability.

The phase-pure QW films deposited with BAAc possess a continuous phase distribution, which not only ensures stability but also guarantees efficient carrier transport along the vertical direction [57].

Clear periodic 2D crystallographic planes (0k0) were observed in all the films (= 2, 3, 4, 5) deposited with BAAc, which further confirmed their high phase purity.

Probing the formation mechanisms of phase-pure QWs

The solution, prepared with stoichiometric amounts of precursors, has been confirmed to be a complex mixture of colloid particles with random size distribution in a mother solution of dissolved ions [50, 51].

This random particle size distribution in the precursor solution is a fundamental limitation for depositing phase-pure QWs films.

In the solution prepared with BAAc, the particles showed a very narrow size distribution (~0.6 nm).

This strong coordination interaction may result in the formation of a stable and homogeneous precursor solution.

For these strong coordination interactions between Ac^- and the precursors as well as the high viscosity and negligible vapour pressure of BAAc (ref [58]), as the other solvents evaporate during the initial stage of spin coating, a uniformly distributed intermediate phase could be gelled from the precursor solution with near-monodisperse unit cell particles.

Photovoltaic performance of solar cells

To demonstrate the potential application of phase-pure QW films, efficient solar cells with remarkably high stability and high V_{OC} were constructed.

The uniform element distribution in the phase-pure QW film is significantly different from the vertical gradient distribution of the elements in the MQW films [34].

With the large increases in V_{OC} and fill factor, the phase-pure QW PSC presented a substantially higher PCE than the MQW PSC.

The phase-pure QW device possesses much better stability than the MQW device under varying conditions.

In phase-pure QW films, the inorganic slabs are uniformly protected by the spacer layers, whereas in the MQW films, there are substantial thick QWs and 3D perovskite crystals.

QW phase purity is of crucial importance for device stability.

Conclusions

Our findings show that the traditional constraints of deposition phase-pure QW films could be overcome by introducing a molten salt spacer.

High-quality phase-pure QW films in 2D Ruddlesden–Popper layered perovskite with microscale vertically aligned grains were produced, yielding an efficient and stable PSC.

Compared to the traditional all-inorganic QWs deposited with vacuum methods [59], the hybrid organic–inorganic metal-halide perovskite phase-pure QWs offer several improvements, including solution processability, low-temperature fabrication and atomic layer accuracy.

Considering the good stability, unique structure and optoelectronic properties, we anticipate that the phase-pure QWs will facilitate the development of solar cells and other perovskite-based optoelectronic devices, such as detectors, light-emitting diodes and lasers.

Methods

The J–V curves were recorded with a Keithley 2400 source meter and 425 W collimated xenon lamp (Enlitech SS-F5-3A) calibrated with the light intensity to 100 mW cm^{-2} under AM1.5G solar light conditions by a certified silicon solar cell (NREL).

The J–V curves were recorded with a Keithley 2400 source meter under one sun illumination (AM1.5G, 100 mW cm^{-2}), equipped with a 425 W collimated xenon lamp (Enlitech SS-F5-3A), in which the light intensity was calibrated by the NREL certified silicon solar cell.

Samples prepared on Si/SnO_2 substrates used an X-ray beam ($\lambda = 0.68877$ Å) at an incident angle of 0.20° for 20 s, and the scattered light was collected by a MarCCD charge-coupled device detector system at a distance of ~746 mm vertically from the sample.

Quantum-Size-Tuned Heterostructures Enable Efficient and Stable Inverted Perovskite Solar Cells

DOI: https://doi.org/10.1038/s41566-022-00985-1

Originators: Chen, H., Teale, S., Chen, B. *et al.* Quantum-size-tuned heterostructures enable efficient and stable inverted perovskite solar cells. *Nat. Photon.* 16, 352–358 (2022). https://doi.org/10.1038/s41566-022-00985-1

Abstract-Summary

The energy landscape of reduced-dimensional perovskites (RDPs) can be tailored by adjusting their layer width (n).

Two/three-dimensional (2D/3D) heterostructures containing n = 1 and 2 RDPs have produced perovskite solar cells (PSCs) with >25% power conversion efficiency (PCE).

We report a method to increase the layer width of RDPs in 2D/3D heterostructures to address this problem.

We discover that bulkier organics form 2D heterostructures more slowly, resulting in wider RDPs; and that small modifications to ligand design induce preferential growth of $n \geq 3$ RDPs.

Main

Studies have revealed that this heterostructure consists of a thin layer of n = 1 and/or n = 2 RDPs atop a bulk 3D perovskite [60–63], which increases performance via passivation and increased electron blocking [64, 65].

Electron blocking is beneficial for devices built in the negative–intrinsic–positive (NIP) architecture, in which the 2D-treated perovskite surface is coated with a hole transport layer (HTL); however, when used in the generally more stable [66–69] inverted (that is, positive–intrinsic–negative, or PIN) structure, this strategy achieves mixed results.

Park and others, La-Placa and others and Zhao and others all found that using a 2D layer in a PIN device reduced the PCE due to a lower current and fill factor [70–72].

We speculated that reducing the confinement in the 2D layer could reduce electron blocking and produce inverted PSCs with an exceptional performance.

A quasi-2D surface treatment

In bulk RDP films, the average layer thickness is modulated by changing the ratio of the 2D ligands to the 3D perovskite precursors in solution [39, 48].

Our most distinctive results came from introducing a small amount of dimethylformamide (DMF) (at 1:200 by volume in isopropanol (IPA)) and methylammonium iodide (MAI) (1:2 by weight) to the 2D ligands in solution.

This earlier study is limited in scope, however, discussing only the effect of DMF on the penetration depth of the 2D treatment.

We instead expect that the origin of improved performance is the production of wider RDPs at the interface between the 3D perovskite and the electron transporting layer (ETL).

Characterizing the quantum well width in 2D/3D heterojunctions

Similar to spectra acquired by Niu and others, the standard 2D-treated films display bleach signals from $n = 1$ and/or 2 RDPs [73].

Quasi-2D-treated films, however, contain $n \geq 3$ RDPs, with the PEA treatment producing more $n \geq 3$ RDPs than BTA.

It was documented that bulk quasi-2D perovskite films made using fluorinated PEA ligands form different QW structures compared to unfluorinated PEA [74].

For the BTA treatment, this reduction in 3D bleach is concurrent with the arrival of a strong $n = 2$ bleach.

For the PEA-based treatments, the 2D signal does not appear until ~1.2 s after the 3D bleach reduction.

We also gained insight into why 3F-PEA forms wider RDPs.

Band alignment of 2D/3D heterostructures

We proposed to modulate the n distribution of 2D/3D heterojunctions with the aim of reducing the electron barrier at the perovskite/ETL interface.

To verify this, we used surface-sensitive inverse photoelectron spectroscopy (IPES), which directly measures the conduction band of the first few nanometres of film [75–79].

Using the quasi-2D treatment the 2D/3D CBM offset is reduced substantially for PEA- and 3F-PEA-treated films relative to their $n = 1$ (and/or 2) analogues (0.42–0.14 eV and 0.49–0.08 eV, respectively) but less so for BTA (0.41–0.31 eV).

This complements the results in the previous section which suggest that the quasi-2D treatment using BTA produces fewer $n \geq 3$ RDPs than PEA or 3F-PEA.

Carrier extraction and photovoltaic performance

To explore the device implications, we investigated the charge transport in BTA-, PEA- and 3F-PEA-treated films using the 2D and quasi-2D treatments.

Using the quasi-2D treatment this was reduced to 4.9 ns, implying a considerably improved extraction compared with the standard 2D treatment.

Repeating these measurements with a quasi-2D-treated film, we observed a decay without PCBM of 5.8 ps, suggesting that the charges are funnelled to the now present $n > 2$ species [39, 80].

We observed a faster decay in the photocurrent for the quasi-2D-treated films ($\tau_{quasi\text{-}2D} = 0.82$ μs compared with $\tau_{2D} = 1.28$ μs for 3F-PEA-treated films), which suggests a substantial improvement in carrier extraction [81].

We found that the n distribution of both treatments was robust under heating, suggesting that the quasi-2D treatment could be used to produce extremely stable devices.

We tracked the maximum power point (MPP) of two quasi-2D-treated devices under illumination at ~50% relative humidity.

Conclusion

We presented a simple technique that alters the QW widths in 2D/3D perovskite heterostructures.

This was tailored to improve inverted PSCs, and was accomplished by reducing the confinement within the 2D capping layer to reduce the electron barrier between the 3D and 2D perovskite species.

This resulted in highly stable NiO_x-based PSCs with certified efficiency values, and could be useful in other metal halide-based optoelectronics devices in which well-passivated, negative–intrinsic contacts are desirable.

Methods

For non-inverted solar cells, the substrate was spin-coated with a thin layer of SnO_2 nanoparticle solution (1:3:3, SnO_2 (15% in water):IPA:water) at 3,000 r.p.m.

for 20 s. During the second spin-coating (2,500 r.p.m.) step, diethyl ether (1 ml) was poured onto the substrate after 15 s. The intermediate phase substrate was then put on a hotplate at 150 °C for 10 min.

After the fabrication of the 3D perovskite film, the 2D layer was fabricated by depositing a solution (5 mg ml^{-1}) of PEAI in IPA onto the perovskite film and then spinning the substrate at 4,000 r.p.m.

for 20 s. During the second spin-coating (2,500 r.p.m.) step, diethyl ether (1 ml) was poured onto the substrate after 15 s. The intermediate phase substrate was then put on a hotplate at 150 °C for 10 min.

Stabilization of 3D/2D Perovskite Heterostructures via Inhibition of Ion Diffusion by Cross-Linked Polymers for Solar Cells with Improved Performance

DOI: https://doi.org/10.1038/s41560-023-01205-y

Originators: Luo, L., Zeng, H., Wang, Z. *et al.* Stabilization of 3D/2D perovskite heterostructures via inhibition of ion diffusion by cross-linked polymers for solar cells with improved performance. *Nat Energy* 8, 294–303 (2023). https://doi.org/10.1038/s41560-023-01205-y

Abstract-Summary

Two-dimensional (2D) and quasi-2D modifications of three-dimensional (3D) perovskite active layers have contributed to advances in the performance of perovskite solar cells (PSCs).

The ionic diffusion between the surface 2D and bulk 3D perovskites leads to the degradation of the 3D/2D perovskite heterostructures and limits the long-term stability of PSCs.

Photoluminescence spectra and thickness-profiled elemental analysis indicate that the CLP stabilizes the heterostructure by inhibiting the diffusion of cations (formamidinium, FA^+ and 4-fluorophenylethylammonium, 4F-PEA^+) between the 2D and 3D perovskites.

Main

The ion diffusion between the surface 2D and bulk 3D perovskites (for example, cations in the 3D perovskite layer move towards the 2D layer) is known to contribute to the degradation of 3D/2D perovskite heterostructures under bias, illumination and heat [82–84].

Inhibiting ion diffusion between the 3D and 2D perovskites and stabilizing the 3D/2D perovskite heterostructures remains an ongoing challenge [82–84].

To construct a stable 3D/2D perovskite heterostructure, we propose to introduce an interlayer between the 3D and 2D perovskite layers.

The interlayer needs to effectively inhibit ion diffusion and, meanwhile, not hinder charge transport between the 3D and 2D perovskite layers.

For the deposition of the interlayer and 2D perovskite layer, the technique and process need to be operated without degrading or damaging the underlying layers.

Comparing small molecules versus regular polymers as the interlayer, we found that a highly cross-linked polymer (CLP) was more effective at inhibiting ion diffusion and stabilizing the 3D/2D perovskite heterostructure.

Selection of CLP

For the design and selection of CLP, it should inhibit the diffusion of FA^+ and 4F-PEA^+ cations between the 3D and 2D perovskite layers and stabilize the 3D/2D perovskite heterostructure.

$PbBr_2$ films dipped in FAI or 4F-PEAI solutions rapidly turn brown or yellow.

This indicates that these small molecules and linear polymers are not ideal candidates for inhibiting the diffusion of FA^+ and 4F-PEA^+ cations.

The $PbBr_2$/pEDMA and $PbBr_2$/CLP films maintained their original colour after dipping in FAI or 4F-PEAI solution.

The XRD patterns show the $PbBr_2$ films were not fully transformed into 3D $FAPb(I_xBr_{1-x})_3$ or 2D 4F-$PEA_2Pb(I_xBr_{1-x})_4$, which indicates that the cross-linked pEDMA and CLP layer can effectively inhibit the diffusion of FA^+ and 4F-PEA^+ cations.

Stabilization effect of CLP on perovskite heterostructures

The XRD spectrum of the aged 3D/CLP/2D perovskite heterostructure shows that the characteristic diffraction peak at 5.2° remains relatively unchanged.

These results indicate that the CLP layer plays the dominant role of stabilizing the 3D/2D perovskite heterostructure under thermal stress.

After the perovskite heterostructures are aged at 100 °C for 120 min, the FA^+ signals largely increase on the surface of the 3D/2D perovskite heterostructure, which indicates the upward diffusion of FA^+ ions and the phase transformation from the 2D perovskite of 4F-PEA_2PbI_4 into 3D perovskite during thermal stress.

The downward diffusion of 4F-PEA^+ ions in the 3D/2D perovskite heterostructure is observed after thermal ageing, which is inhibited by the CLP layer in the 3D/CLP/2D heterostructure.

After ageing at 100 °C for 120 min, the characteristic peak from 4F-PEA_2PbI_4 ($n = 1$) for the 3D/2D heterostructure almost completely vanished, while the peak for the 3D/CLP/2D heterostructure remained unchanged.

Charge carrier dynamics and device performance

The τ_{avg} of the 3D/CLP/2D perovskite heterostructure further increases to 423 ns.

By fitting the linear regions, the electron and hole mobilities of the 3D/CLP/2D heterostructure are similar compared with that of the 3D perovskite or 3D/2D heterostructure.

The average efficiencies of devices based on 3D/2D and 3D/CLP heterostructures are improved to 19.81% and 19.91%, respectively.

The cells based on the 3D/CLP/2D heterostructure deliver an average efficiency of 20.29%.

The efficiency improvement mainly results from the enhanced V_{OC}, which is attributed to the synergistic passivation effect of the CLP and 2D perovskite.

Device stability

The estimated T_{90} lifetime of the 3D/2D PSC is 560 h. The PSC based on the 3D/CLP/2D perovskite heterostructure maintains ~92% of the initial efficiency after 1,000 hours under continuous illumination.

To eliminate the negative effect of spiro-OMeTAD on the cell stability and verify the stability of the perovskite heterostructures, we carry out four repairing cycles of the cell by re-depositing the spiro-OMeTAD layer.

The cell performance recovery indicates that the 3D/CLP/2D perovskite heterostructure is highly stable against light and heat stress.

A more stable poly(3-hexylthiophene)-spiro-OMeTAD (P3HT-spiro-OMeTAD) (3:1 w/w) mixed HTL is adopted to replace the unstable spiro-OMeTAD [85].

Conclusion

We successfully suppress the ion diffusion between surface 2D and bulk 3D perovskites by introducing a highly cross-linked network polymer.

A highly stable 3D/CLP/2D perovskite heterostructure is constructed and combined with a carbon electrode to fabricate PSCs with an efficiency of 21.2% and mini-PSMs with an efficiency of 19.6% (certified stabilized efficiency of 18.2%).

After 4,390 hours of continuous operation under MPP tracking and one-sun illumination, the device retains 90% of its initial performance.

Methods

For the 3D/CLP/2D perovskite film, 3 mg EDMA, POSS (3 mol% of EDMA) and 2,2′-Azobis(2-methylpropionitrile) initiator (2 wt% of EDMA) were first dissolved in 1 ml of chlorobenzene and ethyl acetate mixed solution (v:v = 1:1) to prepare the CLP solution, and then spun on the above 3D perovskite at 4,000 r.p.m.

For fabricating carbon-based modules, laser scribing was performed twice, before and after perovskite film deposition, to form P1 and P2 with 1,064 nm wavelength to complete the series interconnection between sub-cells in the module.

Self-trapped State Enabled Filterless Narrowband Photodetections in 2D Layered Perovskite Single Crystals

DOI: https://doi.org/10.1038/s41467-019-08768-z

This article is licensed under a Creative Commons Attribution 4.0 International License, which permits use, sharing, adaptation, distribution and reproduction in any medium or format, as long as you give appropriate credit to the original author(s) and the source, provide a link to the Creative Commons license, and indicate if changes were made. The images or other third party material in this article are included in the article's Creative Commons license, unless indicated otherwise in a credit line to the material. If material is not included in the article's Creative Commons license and your intended use is not permitted by statutory regulation or exceeds the permitted use, you will need to obtain permission directly from the copyright holder. To view a copy of this license, visit http://creativecommons.org/licenses/by/4.0/.

Originators: Li, J., Wang, J., Ma, J. *et al.* Self-trapped state enabled filterless narrowband photodetections in 2D layered perovskite single crystals. *Nat Commun* 10, 806 (2019). https://doi.org/10.1038/s41467-019-08768-z

Abstract-Summary

Filterless narrowband photodetectors can realize color discrimination without filter or bulk spectrometer, thus greatly reducing the system volume and cost for many imaging applications.

Charge collection narrowing has been demonstrated to be a successful approach to achieve filterless narrowband photodetections; nevertheless, it sacrifices the sensitivity of the photodetectors.

We show a highly tunable narrowband photodetector based on two-dimensional perovskite single crystals with high external quantum efficiency (200%), ultralow dark current (10^{-12} A), and high on–off ratio (10^3).

Introduction

Two-dimensional (2D) Ruddlesden–Popper perovskites have been extensively studied due to the naturally formed multiple quantum-well structure, layered characteristic, flexibly tunable bandgap [14], and the improved long-term stability compared with their three-dimensional (3D) counterparts [11, 31].

The strong electron–phonon coupling leads to the formation of self-trapped states within bandgap in 2D layered perovskites [86].

This enhanced self-trapping in 2D perovskites can alter the electrical and optical properties of the 2D perovskites including the thermal activated broad white light emission [87, 88] and hopping transport of carriers [89].

With these unique features, self-trapped states in 2D perovskites with distinguished weak absorption peak below free exciton absorption naturally form another absorption onset to realize the CCN via recombination losses for narrowband photodetection [90], thus bypassing the necessity to specially design the active materials.

Along with the extremely large anisotropy between the in-plane and out-of-plane electrical conductivity in 2D perovskites [91], we anticipate that the narrowband photodetectors of 2D perovskite single-crystal vertical structures can exhibit excellent performance.

Results

The combination of the absorption engineering and the charge collection manipulation together leads to the narrowband spectral response of the 2D perovskite-based photodetectors [92].

The generic presence of two absorption onsets due to the excitonic absorption and the self-trapped states in $(BA)_2(MA)_{n-1}Pb_nBr_{3n+1}$ ($n = 1$ to 2) and $(BA)_2(MA)_{n-1}Pb_nI_{3n+1}$ ($n = 1$ to 5) series enable us to fabricate narrowband photodetectors based on such 2D perovskite series in the entire visible spectral range.

Further reducing the thickness to 0.3 μm, the X_t peak is no longer observed, as the surface-generated carriers dominate over the bulk-generated carriers due to weak absorption of self-trapped states and thus CCN cannot be achieved, leading to the broadband spectral response and the absence of X_t peak.

The conductivity anisotropy in 2D perovskites can lead to different charge extraction efficiency along the in-plane and out-of-plane direction, and thus the different thickness-dependent narrow spectral response peak position for different device configurations.

Discussion

The presence of self-trapped states can enhance the sub-bandgap absorption leading to the high EQE in our narrowband devices.

We expect that the EQE of 2D perovskite-based devices would be much smaller than that of 3D counterparts if the regular band-tail states lead to the narrowband photoresponse, as the large resistance in the out-of-plane direction makes it more difficult for carriers to be extracted.

We observed an enhancement of EQE in 2D perovskite-based narrowband photodetectors, which should result from the enhanced absorption due to the presence of self-trapped states.

We have demonstrated narrowband photodetections based on 2D perovskite single-crystal plates across the entire visible spectrum with excellent performance by tuning the halide compositions and the layer number n. The highly narrowband spectral response can be attributed to the self-trapped states and extremely large conductivity along the out-of-plane direction.

Methods

The perovskite microplates were exfoliated from respective bulk crystals onto polydimethylsiloxane (PDMS) and an optical microscope coupled with a micromanipulator was used to position the exfoliated microplate onto the predefined electrodes.

The perovskite microplates were exfoliated from the bulk crystals onto PDMS and an optical microscope coupled with a micromanipulator was used to put the exfoliated microplates on the predefined electrodes.

Spray-Coated Perovskite Hemispherical Photodetector Featuring Narrow-Band and Wide-Angle Imaging

DOI: https://doi.org/10.1038/s41467-022-33934-1

This article is licensed under a Creative Commons Attribution 4.0 International License, which permits use, sharing, adaptation, distribution and reproduction in any medium or format, as long as you give appropriate credit to the original author(s) and the source, provide a link to the Creative Commons license, and indicate if changes were made. The images or other third party material in this article are included in the article's Creative Commons license, unless indicated otherwise in a credit line to the material. If material is not included in the article's Creative Commons license and your intended use is not permitted by statutory regulation or exceeds the permitted use, you will need to obtain permission directly from the copyright holder. To view a copy of this license, visit http://creativecommons.org/licenses/by/4.0/.

Originators: Feng, X., He, Y., Qu, W. *et al.* Spray-coated perovskite hemispherical photodetector featuring narrow-band and wide-angle imaging. *Nat Commun* 13, 6106 (2022). https://doi.org/10.1038/s41467-022-33934-1

Abstract-Summary

Sphere imagers featuring specific wavelength recognition and wide-angle imaging are required to meet the fast development of modern technology.

The crystallization speed is manipulated by perovskite compositions, and the film thickness can be controlled by spray-coating cycles and solution concentration from tens of nanometers to hundreds of micrometers with a fast velocity of 1.28×10^{-4} $cm^3\ s^{-1}$.

The lens-free hemispherical photodetectors allow light response at a wide incident angle of 180°.

Wide-angle and wavelength-selective imaging are also demonstrated, which can find potential applications in intelligent recognition and intraoperative navigated surgery.

Introduction

Depositing a soluble active layer to fabricate photosensors on required non-planar substrates to respond to specific wavelength light at a wide angle is still challenging [93, 94].

The low-cost solution processes can deposit layered thin films on desired substrates of irregular shape [95–97].

Spray-coating is a pressure-driven coating process, and sprayed solvents often volatilize very fast under strong airflow before depositing on substrates.

Spray-coated perovskite films for solar cells have been reported for large area devices [98, 99].

We report the spray-coated quasi-2D phenylethylammonium/formamidinium lead halide ($PEA_2FA_{n-1}Pb_nX_{3n+1}$) perovskite hemispherical photodetectors that can collect images upon specific light wavelength at a wide light incident angle of 180°.

The crystallization process of the spray-coated quasi-2D perovskite film is manipulated.

The narrow-band response is realized by controlling the charges carriers' dynamic behavior in a thick hemispherical perovskite device, and wide-angle imaging is also demonstrated.

Results

The hemispherical photodetectors can collect more information regarding different incident directions of light, which is more convenient for wider vision or object positioning.

After each spraying pulse onto substrates, high speed nitrogen (N_2) gas is employed to treat the substrate surfaces and accelerate films crystallization [100–103].

Large-area and uniform perovskite films can be facilely deposited through spray-coating.

We tuned the thickness of the perovskite layer to match the light penetration depth and charge carriers' diffusion length.

The narrow-band response gradually shows up as the thickness of the spray-coated perovskite film increases from 2 μm to 16 μm.

C_{60} is the electron transport layer with a thickness of around 30 nm deposited onto the surface of perovskites.

The hemispherical device shows independent photocurrent upon different light incident angles from 0° to 90°.

The photocurrent of planar device linearly decreases as the incident angle due to the limited light absorption at inclined directions.

Discussion

We report a perovskite hemispherical photodetector through fast spray-coating processes, which performs lens-free imaging capability at a wide-angle range of nearly 180°, greatly reducing the dependency on complex optical components.

The spray-coated perovskite films are large-area, uniform, compact, and compatible to irregularly shaped substrates.

Through manipulating the charges carriers' dynamic behavior in different quasi-2D perovskite compositions and film thicknesses, narrow-band response from visible to near-infrared range is realized on hemispherical perovskite device with FWHMs smaller than 20 nm.

Methods

Substrates were fixed on a thin metal plate and heated to 75 °C before the first layer of the precursor was spray-coated onto them.

After spray-coating, the film is annealed under 100 °C for 10 min.

Films after spray-coating were annealed at 120 °C for 30 min to remove the solvent and crystallization preliminarily.

The perovskite precursor was spray-coated onto the surface of the $PEA_2FA_3PbI_xBr_{13-x}$ at the same condition as a buffer layer.

The volume of the precursor is 800 μL. The temperature of substrates for spray-coating is 75 °C.

After spray-coating, substrates covered by SnO_2 were annealed for 30 min at 170 °C.

The volume of the precursor is 800 uL. The temperature of substrates for spray-coating is 75 °C.

The volume of the precursor is 1600 μL after spray-coating photodetectors were annealed for 15 min at 100 °C.

Bibliography

1. Prajongtat, P., Dittrich, T.: Precipitation of $CH_3NH_3PbCl_3$ in $CH_3NH_3PbI_3$ and its impact on modulated charge separation. J. Phys. Chem. C **119**, 9926–9933 (2015)
2. Jeon, N.J., et al.: Solvent engineering for high-performance inorganic–organic hybrid perovskite solar cells. Nat. Mater. **13**, 897–903 (2014)
3. Li, W., et al.: Enhanced UV-light stability of planar heterojunction perovskite solar cells with caesium bromide interface modification. Energy Environ. Sci. **9**, 490–498 (2016)
4. Wang, Z., et al.: Efficient and air-stable mixed-cation lead mixed-halide perovskite solar cells with n-doped organic electron extraction layers. Adv. Mater. **29**, 1604186 (2016)
5. McMeekin, D.P., et al.: A mixed-cation lead mixed-halide perovskite absorber for tandem solar cells. Science **351**, 151–155 (2016)
6. Bush, K.A., et al.: 23.6%-efficient monolithic perovskite/silicon tandem solar cells with improved stability. Nat. Energy **2**, 17009 (2017)
7. Azpiroz, J.M., Mosconi, E., Bisquert, J., De Angelis, F.: Defect migration in methylammonium lead iodide and its role in perovskite solar cell operation. Energy Environ. Sci. **8**, 2118–2127 (2015)
8. Leguy, A.M., et al.: Reversible hydration of $CH_3NH_3PbI_3$ in films, single crystals, and solar cells. Chem. Mater.Mater. **27**, 3397–3407 (2015)
9. Conings, B., et al.: Intrinsic thermal instability of methylammonium lead trihalide perovskite. Adv. Energy Mater. **5**, 1500477 (2016)
10. Misra, R.K., et al.: Temperature- and component-dependent degradation of perovskite photovoltaic materials under concentrated sunlight. J. Phys. Chem. Lett. **6**, 326–330 (2015)
11. Smith, I.C., Hoke, E.T., Solis-Ibarra, D., McGehee, M.D., Karunadasa, H.I.: A layered hybrid perovskite solar-cell absorber with enhanced moisture stability. Angew. Chem. Int. Ed. **53**(42), 11232–11235 (2014)
12. Yao, K., Wang, X., Xu, Y.X., Li, F., Zhou, L.: Multilayered perovskite materials based on polymeric-ammonium cations for stable large-area solar cell. Chem. Mater. **28**, 3131–3138 (2016)
13. Tsai, H., et al.: High-efficiency two-dimensional Ruddlesden-Popper perovskite solar cells. Nature **536**, 312–316 (2016)
14. Cao, D.H., Stoumpos, C.C., Farha, O.K., Hupp, J.T., Kanatzidis, M.G.: 2D homologous perovskites as light-absorbing materials for solar cell applications. J. Am. Chem. Soc. **137**(24), 7843–7850 (2015). https://doi.org/10.1021/jacs.5b03796
15. Ariga, K., Yamauchi, Y., Mori, T., Hill, J.P.: 25th anniversary article: what can be done with the Langmuir-Blodgett method? Recent developments and its critical role in materials science. Adv. Mater. **25**, 6477–6512 (2013)
16. Lee, J.W., et al.: Formamidinium and cesium hybridization for photo- and moisture-stable perovskite solar cell. Adv. Energy Mater. **5**, 1501310 (2015)
17. Saliba, M., et al.: Incorporation of rubidium cations into perovskite solar cells improves photovoltaic performance. Science **354**, 206–209 (2016)
18. Smith, I. C., Hoke, E. T., Solis-Ibarra, D., McGehee, M. D. & Karunadasa, H. I. A layered hybrid perovskite solar-cell absorber with enhanced moisture stability. Angew. Chem. Int. Ed. **53**, 11232–11235 (2014) (the first research paper on multi-dimensional perovskite (MDP) absorber)
19. Cao, D.H., Stoumpos, C.C., Farha, O.K., Hupp, J.T., Kanatzidis, M.G.: 2D homologous perovskites as light-absorbing materials for solar cell applications. J. Am. Chem. Soc. **137**, 7843–7850 (2015) (in this work, the first systematic work on Class I MDP was presented and the improvement in stability was observed)
20. Stoumpos, C.C., et al.: Ruddlesden–popper hybrid lead iodide perovskite 2D homologous semiconductors. Chem. Mater. **28**, 2852–2867 (2016)
21. Liang, L., Gao, P.: Lead-free hybrid perovskite absorbers for viable application: can we eat the cake and have it too? Adv. Sci. **5**, 1700331 (2018) (this review summarized for the first

time the state-of-the-art development trend of perovskite PV materials and generalized three main motivations behind the recent research interests. The MDP approach is the newest topic in the field)
22. Sichert, J.A., et al.: Quantum size effect in organometal halide perovskite nanoplatelets. Nano Lett. **15**, 6521–6527 (2015)
23. Axtell, E.A., Liao, J.-H., Pikramenou, Z., Kanatzidis, M.G.: Dimensional reduction in II-VI materials: A2Cd3Q4 (A = K, Q = S, Se, Te; A = Rb, Q = S, Se), novel ternary low-dimensional cadmium chalcogenides produced by incorporation of A2Q in CdQ. Chem. A Eur. J. **2**, 656–666 (1996)
24. Axtell, E.A., Park, Y., Chondroudis, K., Kanatzidis, M.G.: Incorporation of A 2 Q into HgQ and dimensional reduction to A2Hg3Q4 and A2Hg6Q7 (A = K, Rb, Cs; Q = S, Se). Access of Li ions in A 2 Hg 6 Q 7 through topotactic ion-exchange. J. Am. Chem. Soc. **120**, 124–136 (1998)
25. Dohner, E.R., Jaffe, A., Bradshaw, L.R., Karunadasa, H.I.: Intrinsic white-light emission from layered hybrid perovskites. J. Am. Chem. Soc. **136**, 13154–13157 (2014)
26. Sourisseau, S., et al.: Reduced band gap hybrid perovskites resulting from combined hydrogen and halogen bonding at the organic−inorganic interface. Chem. Mater. **19**, 600–607 (2007)
27. Calabrese, J., et al.: Preparation and characterization of layered lead halide compounds. J. Am. Chem. Soc. **113**, 2328–2330 (1991)
28. Lin, Y., et al.: Enhanced thermal stability in perovskite solar cells by assembling 2D/3D stacking structures. J. Phys. Chem. Lett. **9**, 654–658 (2018)
29. Yang, W.S., et al.: Iodide management in formamidinium-lead-halide–based perovskite layers for efficient solar cells. Science **356**, 1376–1379 (2017)
30. National Renewable Energy Laboratory: Best Research-Cell Efficiencies (2018). https://www.nrel.gov/pv/assets/pdfs/pv-efficiencies-07-17-2018.pdf
31. Wang, Z., et al.: Efficient ambient-air-stable solar cells with 2D–3D heterostructured butylammonium-caesium-formamidinium lead halide perovskites. Nat. Energy **2**, 17135 (2017)
32. Johnston, N., Howell, L.G.: Sedimentation equilibria of colloidal particles. Phys. Rev. **35**, 274–282 (1930)
33. Dou, L., et al.: Atomically thin two-dimensional organic-inorganic hybrid perovskites. Science **349**, 1518–1521 (2015)
34. Wang, N., et al.: Perovskite light-emitting diodes based on solution-processed self-organized multiple quantum wells. Nat. Photon. **10**, 699 (2016)
35. Yan, J., Qiu, W., Wu, G., Heremans, P., Chen, H.: Recent progress in 2D/quasi-2D layered metal halide perovskites for solar cells. J. Mater. Chem. A **6**, 11063–11077 (2018)
36. Lai, H., Kan, B., Liu, T., Zheng, N., Xie, Z., Zhou, T., Wan, X., Zhang, X., Liu, Y., Chen, Y.: Two-dimensional ruddlesden-popper perovskite with nanorod-like morphology for solar cells with efficiency exceeding 15%. J. Am. Chem. Soc. **140**(37), 11639–11646 (2018). https://doi.org/10.1021/jacs.8b04604
37. Yang, R., et al.: Oriented quasi-2D perovskites for high performance optoelectronic devices. Adv. Mater. **30**, 1804771 (2018)
38. Chondroudis, K., Mitzi, D.B.: Electroluminescence from an organic−inorganic perovskite incorporating a quaterthiophene dye within lead halide perovskite layers. Chem. Mater. **11**(11), 3028–3030 (1999). https://doi.org/10.1021/cm990561t
39. Yuan, M., Quan, L.N., Comin, R., Walters, G., Sabatini, R., Voznyy, O., Hoogland, S., Zhao, Y., Beauregard, E.M., Kanjanaboos, P.: Perovskite energy funnels for efficient light-emitting diodes. Nat. Nanotech. **11**(10), 872–877 (2016). https://doi.org/10.1038/nnano.2016.110
40. Kagan, C., Mitzi, D., Dimitrakopoulos, C.: Organic-inorganic hybrid materials as semiconducting channels in thin-film field-effect transistors. Science **286**(5441), 945–947 (1999). https://doi.org/10.1126/science.286.5441.945
41. Mitzi, D.B., Dimitrakopoulos, C.D., Kosbar, L.L.: Structurally tailored organic−inorganic perovskites: optical properties and solution-processed channel materials for thin-film transistors. Chem. Mater. **13**, 3728–3740 (2001)

42. Long, G., et al.: Spin control in reduced-dimensional chiral perovskites. Nat. Photon. **12**, 528–533 (2018)
43. Zhang, X., et al.: Stable high efficiency two-dimensional perovskite solar cells via cesium doping. Energy Environ. Sci. **10**, 2095–2102 (2017)
44. Zhou, N., Shen, Y., Li, L., Tan, S., Liu, N., Zheng, G., Chen, Q., Zhou, H.: Exploration of crystallization kinetics in quasi two-dimensional perovskite and high performance solar cells. J. Am. Chem. Soc. **140**(1), 459–465 (2018). https://doi.org/10.1021/jacs.7b11157
45. Zhang, X.Q., et al.: Orientation regulation of phenylethylammonium cation based 2D Perovskite solar cell with efficiency higher than 11%. Adv. Energy Mater. **8**, 1702498 (2018)
46. Kamminga, M.E., et al.: Confinement effects in low-dimensional lead iodide perovskite hybrids. Chem. Mater. **28**, 4554–4562 (2016)
47. Quintero-Bermudez, R., et al.: Compositional and orientational control in metal halide perovskites of reduced dimensionality. Nat. Mater. **17**, 900–907 (2018)
48. Quan, L.N., Yuan, M., Comin, R., Voznyy, O., Beauregard, E.M., Hoogland, S., Buin, A., Kirmani, A.R., Zhao, K., Amassian, A., Kim, D.H., Sargent, E.H.: Ligand-stabilized reduced-dimensionality perovskites. J. Am. Chem. Soc. **138**(8), 2649–2655 (2016). https://doi.org/10.1021/jacs.5b11740
49. Tsai, H., et al.: Design principles for electronic charge transport in solution-processed vertically stacked 2D perovskite quantum wells. Nat. Commun. **9**, 2130 (2018)
50. Yan, K., et al.: Hybrid halide perovskite solar cell precursors: colloidal chemistry and coordination engineering behind device processing for high efficiency. J. Am. Chem. Soc. **137**, 4460–4468 (2015)
51. McMeekin, D.P., et al.: Crystallization kinetics and morphology control of formamidinium–cesium mixed-cation lead mixed-halide perovskite via tunability of the colloidal precursor solution. Adv. Mater. **29**, 1607039 (2017)
52. Proppe, A.H., Quintero-Bermudez, R., Tan, H., Voznyy, O., Kelley, S.O., Sargent, E.H.: Synthetic control over quantum well width distribution and carrier migration in low-dimensional perovskite photovoltaics. J. Am. Chem. Soc. **140**(8), 2890–2896 (2018). https://doi.org/10.1021/jacs.7b12551
53. Liu, J., Leng, J., Wu, K., Zhang, J., Jin, S.: Observation of internal photoinduced electron and hole separation in hybrid two-dimentional perovskite films. J. Am. Chem. Soc. **139**(4), 1432–1435 (2017). https://doi.org/10.1021/jacs.6b12581
54. Liang, Y., et al.: Lasing from mechanically exfoliated 2D homologous Ruddlesden-Popper perovskite engineered by inorganic layer thickness. Adv. Mater. **39**, 1903030 (2019)
55. Yantara, N., et al.: Designing efficient energy funneling kinetics in Ruddlesden-Popper perovskites for high-performance light-emitting diodes. Adv. Mater. **30**, 1800818 (2018)
56. Byun, J., et al.: Efficient visible quasi-2D perovskite light-emitting diodes. Adv. Mater. **28**, 7515–7520 (2016)
57. Xing, G., et al.: Transcending the slow bimolecular recombination in lead-halide perovskites for electroluminescence. Nat. Commun. **8**, 14558 (2017)
58. Chao, L., et al.: Room-temperature molten salt for facile fabrication of efficient and stable perovskite solar cells in ambient air. Chem **5**, 995–1006 (2019)
59. Okamoto, K., et al.: Surface-plasmon-enhanced light emitters based on InGaN quantum wells. Nat. Mater. **3**, 601–605 (2004)
60. Teale, S., et al.: Dimensional mixing increases the efficiency of 2D/3D perovskite solar cells. J. Phys. Chem. Lett. **11**, 5115–5119 (2020)
61. Yoo, J.J., et al.: An interface stabilized perovskite solar cell with high stabilized efficiency and low voltage loss. Energy Environ. Sci. **12**, 2192–2199 (2019)
62. Hu, Y., et al.: Hybrid perovskite/perovskite heterojunction solar cells. ACS Nano **10**, 5999–6007 (2016)
63. Proppe, A.H., et al.: Multication perovskite 2D/3D interfaces form via progressive dimensional reduction. Nat. Commun. **12**, 3472 (2021)
64. Jang, Y.W., et al.: Intact 2D/3D halide junction perovskite solar cells via solid-phase in-plane growth. Nat. Energy **6**, 63–71 (2021)

65. Zhang, T., et al.: Stable and efficient 3D–2D perovskite-perovskite planar heterojunction solar cell without organic hole transport layer. Joule **2**, 2706–2721 (2018)
66. Bai, S., et al.: Planar perovskite solar cells with long-term stability using ionic liquid additives. Nature **571**, 245–250 (2019)
67. Lin, Y.H., et al.: A piperidinium salt stabilizes efficient metal-halide perovskite solar cells. Science **369**, 96–102 (2020)
68. Zheng, X., et al.: Managing grains and interfaces via ligand anchoring enables 22.3%-efficiency inverted perovskite solar cells. Nat. Energy **5**, 131–140 (2020)
69. Chen, S., Xiao, X., Gu, H., Huang, J.: Iodine reduction for reproducible and high-performance perovskite solar cells and modules. Sci. Adv. **7**, eabe8130 (2021)
70. Park, S.M., Abtahi, A., Boehm, A.M., Graham, K.R.: Surface ligands for methylammonium lead iodide films: surface coverage, energetics, and photovoltaic performance. ACS Energy Lett. **5**, 799–806 (2020)
71. La-Placa, M.G., et al.: Vacuum-deposited 2D/3D perovskite heterojunctions. ACS Energy Lett. **4**, 2893–2901 (2019)
72. Zhao, T., Chueh, C.-C., Chen, Q., Rajagopal, A., Jen, A.K.-Y.: Defect passivation of organic–inorganic hybrid perovskites by diammonium iodide toward high-performance photovoltaic devices. ACS Energy Lett. **1**, 757–763 (2016)
73. Niu, T., et al.: Interfacial engineering at the 2D/3D heterojunction for high-performance perovskite solar cells. Nano Lett. **19**, 7181–7190 (2019)
74. Hu, J., Oswald, I.W.H., Stuard, S.J., Nahid, M.M., Zhou, N., Williams, O.F., Guo, Z., Yan, L., Hu, H., Chen, Z., Xiao, X., Lin, Y., Yang, Z., Huang, J., Moran, A.M., Ade, H., Neilson, J.R., You, W.: Synthetic control over orientational degeneracy of spacer cations enhances solar cell efficiency in two-dimensional perovskites. Nat. Commun. **10**(1), 1276 (2019). https://doi.org/10.1038/s41467-019-08980-x
75. Mahmud, M.A., et al.: Double-sided surface passivation of 3D perovskite film for high-efficiency mixed-dimensional perovskite solar cells. Adv. Funct. Mater. **30**, 1907962 (2020)
76. Briggs, D., Seah, M.P. (eds.).: Practical Surface Analysis. Auger and X-ray Photoelectron Spectroscopy, 2nd edn, vol. 1. Wiley (1990)
77. Boehm, A.M., Liu, T., Park, S.M., Abtahi, A., Graham, K.R.: Influence of surface ligands on energetics at FASnI3/C60 interfaces and their impact on photovoltaic performance. ACS Appl. Mater. Inter. **12**, 5209–5218 (2020)
78. Yoshida, H.: Near-ultraviolet inverse photoemission spectroscopy using ultra-low energy electrons. Chem. Phys. Lett. **539–540**, 180–185 (2012)
79. Boehm, A.M., Wieser, J., Butrouna, K., Graham, K.R.: A new photon source for ultraviolet photoelectron spectroscopy of organic and other damage-prone materials. Org. Electron. **41**, 9–16 (2017)
80. Proppe, A.H., et al.: Spectrally resolved ultrafast exciton transfer in mixed perovskite quantum wells. J. Phys. Chem. Lett. **10**, 419–426 (2019)
81. Chen, W., et al.: Efficient and stable large-area perovskite solar cells with inorganic charge extraction layers. Science **350**, 944–948 (2015)
82. Sidhik, S., et al.: Deterministic fabrication of 3D/2D perovskite bilayer stacks for durable and efficient solar cells. Science **377**, 1425–1430 (2022)
83. Sutanto, A.A., et al.: In situ analysis reveals the role of 2D perovskite in preventing thermal-induced degradation in 2D/3D perovskite interfaces. Nano Lett. **20**, 3992–3998 (2020)
84. Sutanto, A.A., et al.: Dynamical evolution of the 2D/3D interface: a hidden driver behind perovskite solar cell instability. J. Mater. Chem. A **8**, 2343–2348 (2020)
85. Liu, Z., et al.: A holistic approach to interface stabilization for efficient perovskite solar modules with over 2,000-hour operational stability. Nat. Energy **5**, 596–604 (2020)
86. Thirumal, K., et al.: Morphology-independent stable white-light emission from self-assembled two-dimensional perovskites driven by strong exciton–phonon coupling to the organic framework. Chem. Mater. **29**, 3947–3953 (2017)
87. Dohner, E.R., Jaffe, A., Bradshaw, L.R., Karunadasa, H.I.: Intrinsic white-light emission from layered hybrid perovskites. J. Am. Chem. Soc. **136**(38), 13154–13157 (2014). https://doi.org/10.1021/ja507086b

88. Hu, T., et al.: Mechanism for broadband white-light emission from two-dimensional (110) hybrid perovskites. J. Phys. Chem. Lett. **7**, 2258–2263 (2016)
89. Straus, D.B., Kagan, C.R.: Electrons, excitons, and phonons in two-dimensional hybrid perovskites: connecting structural, optical, and electronic properties. J. Phys. Chem. Lett. **9**(6), 1434–1447 (2018). https://doi.org/10.1021/acs.jpclett.8b00201
90. Wu, X.X., et al.: Trap states in lead iodide perovskites. J. Am. Chem. Soc. **135**, 2089–2096 (2015)
91. Wang, K., Wu, C., Yang, D., Jiang, Y., Priya, S.: Quasi-two-dimensional halide perovskite single crystal photodetector. ACS Nano **12**(5), 4919–4929 (2018). https://doi.org/10.1021/acsnano.8b01999
92. Fang, Y., Dong, Q., Shao, Y., Yuan, Y., Huang, J.: Highly narrowband perovskite single-crystal photodetectors enabled by surface-charge recombination. Nat. Photon. **9**, 679–686 (2015)
93. Kogos, L.C., et al.: Plasmonic ommatidia for lensless compound-eye vision. Nat. Commun. **11**, 1637 (2020)
94. Wang, J., et al.: Self-driven perovskite narrowband photodetectors with tunable spectral responses. Adv. Mater. **33**, e2005557 (2021)
95. García de Arquer, F.P., Armin, A., Meredith, P., Sargent, E.H.: Solution-processed semiconductors for next-generation photodetectors. Nat. Rev. Mater. **2**, 16100 (2017)
96. Bishop, J.E., et al.: High-efficiency spray-coated perovskite solar cells utilizing vacuum-assisted solution processing. ACS Appl. Mater. Inter. **10**, 39428–39434 (2018)
97. Hong, S.C., et al.: Precise morphology control and continuous fabrication of perovskite solar cells using droplet-controllable electrospray coating system. ACS Appl. Mater. Inter. **9**, 7879–7884 (2017)
98. Heo, J.H., Lee, M.H., Jang, M.H., Im, S.H.: Highly efficient $CH_3NH_3PbI_{3-x}Cl_x$ mixed halide perovskite solar cells prepared by re-dissolution and crystal grain growth via spray coating. J. Mater. Chem. A **4**, 17636–17642 (2016)
99. Park, M., et al.: Highly reproducible large-area perovskite solar cell fabrication via continuous megasonic spray coating of $CH_3NH_3PbI_3$. Small **15**, e1804005 (2019)
100. Wang, J., et al.: Templated growth of oriented layered hybrid perovskites on 3D-like perovskites. Nat. Commun. **11**, 582 (2020)
101. Deng, Y., et al.: Surfactant-controlled ink drying enables high-speed deposition of perovskite films for efficient photovoltaic modules. Nat. Energy **3**, 560–566 (2018)
102. Deng, Y., et al.: Tailoring solvent coordination for high-speed, room-temperature blading of perovskite photovoltaic films. Sci. Adv. **5**, eaax7537 (2019)
103. Cassella, E.J., et al.: Gas-assisted spray coating of perovskite solar cells incorporating sprayed self-assembled monolayers. Adv. Sci. **9**, e2104848 (2022)

Chapter 5
Applications II: Light-Emitting Diodes

Introduction by Editor

The next part of the story of optical applications for metal halide perovskites is that the very property which can give 2D MHPs a disadvantage over 3D MHPs for photovoltaic applications, namely the quantum and dielectric confinement which leads to a very large exciton binding energy and makes it harder to separate the photo-excited electrons and holes, can give 2D MHPS a potentially large advantage in light emitting applications. The natural quantum well structures which confine carriers and increase the exciton binding energies, also help the electrons and holes recombine very efficiently. The other advantages of 2D MHPs are also applicable to LEDs, namely great tunability in bandgaps, defect tolerance relative to III-V semiconductors, and improved stability relative to 3D MHPS. Therefore, research into 2D MHP LEDs is being very aggressively pursued. In this section we have included 13 papers on the study and development of LEDs, two of which are themselves review papers which summarize a number of additional developments (Zhang, et al., 2021, and Ren, et al., 2021).

A main theme of these papers is the ongoing quest for increased efficiency. Nonradiative Auger recombination, in which the energy from electron-hole recombination is used to excite other electrons in the conduction band into higher CB states (or holes in the valence band to lower VB states) instead of creating photons, is known to be a significant source of inefficiency. Auger recombination has a strong dependence on carrier concentration, so one method to reduce it includes using larger n numbers (number of metal halide monolayers within each inorganic layer) to form wider wells. Other known sources of inefficiency include surface, interface, and bulk defects (e.g. uncoordinated Pb^{2+} ions). In many cases specialized organic molecules can be used to passivate these defects, and several of these papers focus on such

© The Author(s), under exclusive license to Springer Nature Singapore Pte Ltd. 2024

J. S. Colton and K. R. Hansen (eds.), *Two-Dimensional Metal Halide Perovskites*,
https://doi.org/10.1007/978-981-99-7830-4_5

developments. Properly chosen organic molecules can also suppress ion transport with its accompanying degradation of the material.

N number engineering can also be used to funnel electrons and holes from layers with higher bandgaps (low n number) to layers with lower bandgaps (higher n numbers). This type of energy transfer among different n number layers can help avoid concentration quenching in the emitting layer. A similar principle has also been used to funnel carriers into MHP quantum dots from a higher bandgap 2D MHP surrounding matrix, one of the tricks used to obtain higher quality blue LEDs, which have proven more elusive than lower wavelength LEDs. On the other hand, severe phase segregation can lead to device failure and many researchers are therefore focusing on obtaining more uniform n numbers in order to obtain longer device lifetimes. More uniform n numbers also leads to greater uniformity in emitted wavelengths, which then leads to better color purity for devices.

These examples illustrate the progress that has been made and the work there is yet to be done. The great promise of these materials—namely high performance LEDs with low-cost fabrication, great spectral tunability, potential white light sources, possible printable and flexible light emitters, and also a potential for reduced toxicity (relative to 3D MHPs)—ensures that the study and development of these materials for light emitting applications will continue strongly into the future.

Machine generated summaries

Disclaimer: The auto-summaries have been generated by a recursive clustering algorithm via the Dimensions Auto-summarizer by Digital Science handled by Subject Matter Experts and the external editor. The editor of this book selected which SN content should be auto-summarized and decided its order of appearance. Please be aware that the auto-summaries consist of original sentences, but are not representative of its original paper, since we do not show the full length of the publication. Please note that only published SN content is represented here, and that machine-generated books are still at an experimental stage.

Machine generated keywords: quasid, peled, led, quasid perovskite, blue, electroluminescence, perovskite lightemitte, perovskite led, lighting, quasitwodimensional, green, eqe, quenching, emission, qws

Perovskite Energy Funnels for Efficient Light-Emitting Diodes

DOI: https://doi.org/10.1038/nnano.2016.110

Originators: Yuan, M., Quan, L., Comin, R. *et al.* Perovskite energy funnels for efficient light-emitting diodes. *Nature Nanotech* 11, 872–877 (2016). https://doi.org/10.1038/nnano.2016.110

Abstract-Summary

Organometal halide perovskites exhibit large bulk crystal domain sizes, rare traps, excellent mobilities and carriers that are free at room temperature—properties that support their excellent performance in charge-separating devices.

In devices that rely on the forward injection of electrons and holes, such as light-emitting diodes (LEDs), excellent mobilities contribute to the efficient capture of non-equilibrium charge carriers by rare non-radiative centres.

The materials function as charge carrier concentrators, ensuring that radiative recombination successfully outcompetes trapping and hence non-radiative recombination.

We use the new material to build devices that exhibit an external quantum efficiency (EQE) of 8.8% and a radiance of 80 W sr^{-1} m^{-2}.

Main

The photoluminescence quantum yield (PLQY) of $CH_3NH_3PbI_3$ perovskites therefore depends on the excitation intensity, and reaches a higher value at high excitation photon fluences where radiative bimolecular recombination dominates.

The performance of perovskite LEDs has been severely curtailed by the low radiative efficiency of the active medium under device operating conditions.

To improve LED performance, perovskite films are typically engineered to be thin (~20 nm) to both confine the injected charges and maximize the carrier density [1].

Further improving the efficiency of radiative recombination under low-level injection is an important route to more efficient perovskite LEDs [2].

Two-dimensional (2D) layered perovskites [3, 4] provide one means to form excitons at room temperature, with the largest exciton binding energy reaching an impressive $E_b \approx 320$ meV (ref.

We sought to explore whether increasing the average number of layers $\langle n \rangle$ that make up the average dimensionally tuned perovskite in a solid-state material could be used as a means to enhance the PLQY. ⟨

Structure and photoluminescence of quasi-2D perovskites

In the case of quasi-2D perovskites, a series of Bragg reflections at low angles ($2\theta < 10°$) was observed, revealing the vertical growth of the unit cell, which is indicative of lead iodide $\langle n \rangle$-layer perovskite sheets [5].

X-ray photoelectron spectroscopy (XPS) was used to determine the stoichiometry present at the surface of quasi-2D perovskite films.

The 3D $CH_3NH_3PbI_3$ perovskite films exhibits low PLQY (~0.2%), a fact that we attribute to a dominance of trap-mediated non-radiative recombination, consistent with previous reports [6].

The 2D perovskite compound, PEA_2PbI_4 (n = 1), also exhibits low photoluminescence in films (PLQY <0.1%), again in agreement with previous studies [3, 7].

Quasi-2D perovskites with $\langle n \rangle = 3$ and $\langle n \rangle = 5$ show a substantially more intense photoluminescence: their respective PLQY values reach 10.1 and 10.6% under the same low (6 mW cm^{-2}) excitation intensities for which the PLQY of 3D materials is negligible.

Transient dynamics

These results suggest that the quasi-2D perovskite films are not single-phase, but rather consist of a collection of grains exhibiting a variety of ⟨n⟩ values.

The fact that the steady-state PL spectrum of the ⟨n⟩ = 3 perovskite film reflects the smallest bandgap component already indicates that photoexcitation transfer is at play in these materials.

Similar phenomena were observed in the ⟨n⟩ = 5 quasi-2D perovskite films, where an inhomogeneously broadened bleaching peak was observed around 700 nm.

These studies of excitation dynamics suggest that the multiphased perovskite materials $PEA_2(MAI)_{n-1}Pb_nI_{3n+1}$ channel energy across an inhomogeneous energy landscape, concentrating carriers on smaller bandgap emitters.

We posit that the high PLQY under low excitation is achieved because only the subset of trap states found inside the lowest-energy inclusions need to be filled and there is an increased local excitation intensity within small bandgap grains, which is otherwise achievable only at higher excitation intensities in 3D perovskites.

We then investigated the film morphologies of quasi-2D perovskites with different ⟨n⟩ values.

LED structure and performance

By taking advantage of the solvent engineering fabrication process [8], we were able to achieve high-quality, ultra-smooth perovskite layers without pinholes for different ⟨n⟩ values.

These results represent the most efficient and brightest solution-processed LED operating at near-infrared wavelengths reported so far, with an ~2.4× increase in peak EQE and an ~3× increase in peak radiance over the previous best-reported near-infrared perovskite EL devices.

The EQE in ⟨n⟩ = 5 perovskite films rises from 0.14% for 50 nm to 8.8% for 200 nm films, whereas the radiance increases from 1.2 to 80 W sr^{-1} m^{-2}.

The competition between non-radiative and radiative recombination governs the evolution of the EQE across perovskite devices with different ⟨n⟩ values.

Conclusions

The low efficiency of radiative recombination in EL devices based on $CH_3NH_3PbI_3$ perovskites motivated our exploration of new strategies to overcome this limitation.

We found that multicomponent multilayered perovskite solids exhibit record-high PLQY under low excitation fluences.

Dynamic studies enabled us to attribute this advance to the conveyance of excited carriers in a funnelling mechanism that provides carrier concentration [9, 10], enabling more efficient radiative recombination and thereby overcoming the trap-mediated non-radiative recombination at much more practical fluences.

Methods

The following settings were used for all quantum efficiency measurements: an excitation wavelength of 440 nm, bandpass values of 10 and 5 nm for the excitation and emission slits, respectively, step increments of 1 nm and integration times of 0.5 s per data point.

The EQE was determined as the ratio, per unit time, of the number of emitted photons to the number of injected electrons [11, 12]: where, I(V) is the current passing through the device at an applied bias, V, $N_{phot(V)}$ represents the number of emitted photons collected by the photodiode and the geometric factor, g, is the solid angle of the EL profile, assumed to be Lambertian, subtended by the photodiode, $\Omega = \pi/g$: where a is the diameter of the active area of the photodiode and L represent the distance between the emitting pixel and the active area.

The PL intensity was measured in temperature steps of 20 K at the same spatial location on the sample.

Perovskite Light-Emitting Diodes Based on Solution-Processed Self-organized Multiple Quantum Wells

DOI: https://doi.org/10.1038/nphoton.2016.185

Originators: Wang, N., Cheng, L., Ge, R. *et al.* Perovskite light-emitting diodes based on solution-processed self-organized multiple quantum wells. *Nature Photon* 10, 699–704 (2016). https://doi.org/10.1038/nphoton.2016.185

Main

Previous studies on PbI_4-based perovskites showed that the absorption peaks at around 2.4, 2.2, 2.0 and 1.9 eV correspond to excitonic absorption of QWs with n = 1, 2, 3 and 4 respectively [13–15].

The high PLQE of up to 60% suggests that the photogenerated excitons in small-n QWs can be energy transferred to the large-n QWs with superior emission properties, allowing efficient radiative recombination.

In EL device operation, electrons and holes are injected into the perovskite layer and accumulated at the large-n QWs because of the cascade energy structure of the perovskite MQW films.

To fabricate high-efficiency perovskite LEDs, it is a challenge to obtain a large-n emissive thin layer with good coverage sandwiched between two non-quenching contacts with appropriate energy levels, while our MQW structure provides a simple alternative solution to it.

Methods

The perovskite films were prepared by spin-coating the precursor solution onto the PEIE treated ZnO films, followed by annealing on a hot plate at 100 °C.

PL spectra of the perovskite films were measured at room temperature using a fluorescent spectrophotometer (F-4600, HITACHI) with a 200 W Xe lamp as an excitation source.

The excitation spectra and time-resolved fluorescence spectra of the perovskite films were obtained by using an Edinburgh Instruments (FLS920) spectrometer.

For the time-resolved PL measurements, the perovskite films were excited from the quartz glass substrate side by a 405 nm pulsed diode laser (EPL-405) with a fluence of around 4 nJ cm^{-2}.

A three-step technique was used to obtain the PLQE of perovskite films by combination of a 445 nm continuous wave laser, optical fibre, spectrometer and integrating sphere [16].

The thickness of the perovskite films is about 30 nm (same as in EL devices) for the above optical characterizations.

Efficient Green Light-Emitting Diodes Based on Quasi-Two-Dimensional Composition and Phase Engineered Perovskite with Surface Passivation

DOI: https://doi.org/10.1038/s41467-018-02978-7

This article is licensed under a Creative Commons Attribution 4.0 International License, which permits use, sharing, adaptation, distribution and reproduction in any medium or format, as long as you give appropriate credit to the original author(s) and the source, provide a link to the Creative Commons license, and indicate if changes were made. The images or other third party material in this article are included in the article's Creative Commons license, unless indicated otherwise in a credit line to the material. If material is not included in the article's Creative Commons license and your intended use is not permitted by statutory regulation or exceeds the permitted use, you will need to obtain permission directly from the copyright holder. To view a copy of this license, visit http://creativecommons.org/licenses/by/4.0/.

Originators: Yang, X., Zhang, X., Deng, J. *et al.* Efficient green light-emitting diodes based on quasi-two-dimensional composition and phase engineered perovskite with surface passivation. *Nat Commun* 9, 570 (2018). https://doi.org/10.1038/s41467-018-02978-7

Abstract-Summary

Perovskite light-emitting diodes (LEDs) are attracting great attention due to their efficient and narrow emission.

Quasi-two-dimensional perovskites with Ruddlesden–Popper-type layered structures can enlarge exciton binding energy and confine charge carriers and are considered good candidate materials for efficient LEDs.

Firstly, through composition and phase engineering, optimal quasi-two-dimensional perovskites are selected.

Green LEDs based on quasi-two-dimensional perovskite reach a current efficiency of 62.4 cd A^{-1} and external quantum efficiency of 14.36%.

Introduction

The exciton binding energy is enlarged in the two-dimensional (2D) perovskite and thus the emission efficiency can be enhanced [7, 17].

This makes the quasi-2D perovskite an efficient luminescent material [18–21].

The EQEs of quasi-2D perovskite LEDs have achieved around 10% [18, 19] and highest EQE of 11.7% in near-infrared region was obtained [19].

That quasi-2D perovskite materials usually contain a mixture of phases and the phase impurity could lead to low emission efficiency, and thus fine control of the composition/phases is critical for efficient emission.

Solution processed thin films of quasi-2D perovskite show reduced crystal size compared to those of 3D perovskite [18, 20].

$PEA_2(FAPbBr_3)_{n-1}PbBr_4$ ($n = 1, 2, \ldots, \infty$) is investigated as the emitting layer for quasi-2D perovskite LEDs, where PEA and FA are phenylethylammonium ($C_6H_5C_2H_4NH_3^+$) and formamidinium ($HC(NH_2)_2^+$), respectively.

We present surface passivation of quasi-2D perovskite films by trioctylphosphine oxide (TOPO) treatment [22].

Results

Previous results suggested that the quasi-2D perovskite films are not a single phase but contain a variety of n-phases [18, 19].

This result shows that all the quasi-2D perovskite films always contain large n-phases ($n \rightarrow \infty$) or 3D perovskite regardless of the composition of the precursor.

The 3D perovskite showed a band edge emission peak at 542 nm (2.29 eV) [23].

All quasi-2D perovskite films have two main emission peaks: one in the green region (approximately 532 nm, approximately 2.33 eV) and another in the blue region (at 440 nm, 2.82 eV).

There are some exceptions: for the $n = 2$ composition, we have not observed the $n = 1$ phase emission, while the $n = 4$ composition shows an emission at 405 nm from the 2D perovskite phase ($n = 1$), and the reason for this is not clear at present.

Discussion

Our device showed the almost same level stability of I-related perovskite with the stability in several 100 min [20].

Compared with the Br-related perovskite with several 10 min stability in previous reports [20, 24], our devices showed a little bit improvement.

One major issue could be ion migration under voltage bias, and another issue could be the phase stability of perovskite materials under heating or moist conditions.

Several strategies could be adopted to improve PeLED stability such as finding stable perovskite-emitting materials [25, 26], incorporating suitable groups to avoid or suppress the ion movements [25, 27], seeking effective ligands [28, 29] to bond with the grain surface and also the film surface to reduce Joule heating induced by non-radiative recombination and finding stable charge injection layers [30].

Methods

For surface passivation, 2 mg ml^{-1} TOPO in CB was dripped onto the perovskite film for a 1 min to let TOPO passivate the film surface enough and then spin-coated at 7000 rpm for 1 min.

XRD measurements were performed with a Rigaku D/max 2500H equipment with a conventional Cu target X-ray tube (Cu K-alpha, $\lambda = 1.5418$ Å) set to 40 kV and 200 mA. Steady-state PL spectra and time-resolved fluorescence spectra (TRPL) of the perovskite films were measured at room temperature in the ambient air using a fluorescent spectrophotometer (HORIBA JY Nanolog-TCSPC).

The PLQY of the perovskite films was measured using an Edinburgh FLS920 equipment with a 450 W xenon lamp as excitation source.

The following settings were applied for PLQY measurements: excitation wavelength of 400 nm; bandpass values of 3.00 and 0.35 nm for the excitation and emission slits, respectively; step increments of 0.5 nm and integration time of 0.3 s per data point.

Minimising Efficiency Roll-Off in High-Brightness Perovskite Light-Emitting Diodes

DOI: https://doi.org/10.1038/s41467-018-03049-7

This article is licensed under a Creative Commons Attribution 4.0 International License, which permits use, sharing, adaptation, distribution and reproduction in any medium or format, as long as you give appropriate credit to the original author(s) and the source, provide a link to the Creative Commons license, and indicate if changes were made. The images or other third party material in this article are included in the article's Creative Commons license, unless indicated otherwise in a credit line to the material. If material is not included in the article's Creative Commons license and your intended use is not permitted by statutory regulation or exceeds the permitted use, you will need to obtain permission directly from the copyright holder. To view a copy of this license, visit http://creativecommons.org/licenses/by/4.0/.

Originators: Zou, W., Li, R., Zhang, S. *et al.* Minimising efficiency roll-off in high-brightness perovskite light-emitting diodes. *Nat Commun* 9, 608 (2018). https://doi.org/10.1038/s41467-018-03049-7

Abstract-Summary

We present investigations of the efficiency roll-off in perovskite LEDs based on two-dimensional layered perovskites.

By simultaneously measuring electroluminescence and photoluminescence on a working device, supported by transient photoluminescence decay measurements, we conclude that the efficiency roll-off in perovskite LEDs is mainly due to luminescence quenching which is likely caused by non-radiative Auger recombination.

This approach leads to the realization of a perovskite LED with a record external quantum efficiency of 12.7%, and the efficiency remains to be high, at approximately 10%, under a high current density of 500 mA cm^{-2}.

Introduction

Low-temperature solution-processed metal halide perovskites show excellent luminescence and charge transport properties, high color purity, and tunable bandgap, making them promising for low-cost and high-performance light-emitting diode (LED) applications [1, 18, 19, 31].

Compared to other similar technologies, such as organic LEDs and quantum-dot LEDs, perovskite LEDs can potentially achieve better efficiency at high brightness since its charge mobility is higher and luminescence quenching due to high exciton concentration is less significant [31, 32].

High external quantum efficiency (EQE) up to 11.7% can be obtained with the MQW perovskites based LED because the lower energy gap QWs which generate electroluminescence (EL) are confined by higher bandgap QWs [19].

We find that the efficiency roll-off in PeLEDs is mainly due to luminescence quenching most likely caused by non-radiative Auger recombination, and the luminescence quenching effect can be suppressed by increasing the width of perovskite QWs.

Results

In order to investigate which mechanism dominates the efficiency roll-off in the MQW perovskite LEDs, we measure the PLQE and EQE of the devices simultaneously by using low-frequency chopped illumination at very low intensity (0.03 mW cm^{-2}), when the LED device is in operation.

The quenched PL intensities (and hence the efficiency roll-off) at high intensities could result from either high carrier (and/or exciton) density or high electrical field across the emissive layer [33–35].

In order to disentangle these two different mechanisms, we monitor the PL response at negative bias, where the electric field is high and the carrier/exciton density is low.

These results indicate that at low carrier densities produced by photoexcitation, the internal electrical field in the device quenches the luminescence of MQW perovskites.

If the efficiency roll-off in our MQW perovskite LEDs is caused by an Auger recombination, then it can be improved by reducing the local carrier density in QWs.

Discussion

By simultaneously measuring the EL and PL on a working device, supported by transient PL decay measurements, we conclude that the efficiency roll-off in perovskite LEDs is mainly due to luminescence quenching which is most likely caused by an Auger recombination.

In GaN-based LEDs, they usually require sophisticated control to reduce the luminescence quenching [36].

While in the solution-processed MQW perovskite LEDs, the processing simplicity of the MQW structures allows easy tunability of the QW width, resulting in reduced luminescence quenching and enhanced device performance.

Our work provides a guide to achieve high-efficiency, high-brightness, and low-roll-off perovskite LEDs, which may also have implications for the development of other kinds of LEDs.

Methods

The devices were fabricated on a glass substrate coated with indium tin oxide and has the structure: ITO/polyethylenimine ethoxylated (PEIE)-modified ZnO (20 nm)/perovskite (30 nm)/poly(9,9-dioctyl-fluorene-co-N-(4-butylphenyl)diphenylamine) (TFB, 40 nm)/MoO_3 (7 nm)/Au (60 nm).

The EL devices were characterized on top of the integration sphere, and notably, only forward light emission could be collected.

Steady-state PL spectra of the perovskite films were recorded by using a fluorescent spectrophotometer (F-4600, HITACHI) with a 200 W Xe lamp as an excitation source.

Transient PL spectra and the PL decay were obtained by using an Edinburgh Instruments spectrometer (FLS980) with a 445-nm pulsed laser (less than 100 ps, 1 MHz, YSL Supercontinuum Source SC-PRO).

For the PLQE measurement, a three-step technique was used through the combination of a 445-nm CW laser, an optical fiber, a spectrometer, and an integrating sphere [16].

The LED devices were excited by a cw 445 nm laser with an optical chopper for frequency modulation (930 Hz).

Efficient Blue Light-Emitting Diodes Based on Quantum-Confined Bromide Perovskite Nanostructures

DOI: https://doi.org/10.1038/s41566-019-0505-4

Originators: Liu, Y., Cui, J., Du, K. *et al.* Efficient blue light-emitting diodes based on quantum-confined bromide perovskite nanostructures. *Nat. Photonics* 13, 760–764 (2019). https://doi.org/10.1038/s41566-019-0505-4

Abstract-Summary

The performance of blue perovskite LEDs (PeLEDs) [37–46] is far inferior to that of their near-infrared, red and green counterparts [18–20, 47–51], strongly limiting the practicality of the PeLED technology.

We demonstrate blue PeLEDs emitting at 483 nm with colour coordinates of (0.094, 0.184) and operating with a peak external quantum efficiency of up to 9.5% at a luminance of 54 cd m^{-2}.

The efficient blue electroluminescence originates from a structure of quantum-confined perovskite nanoparticles embedded within quasi-two-dimensional phases with higher bandgaps, prepared by an antisolvent processing scheme.

Main

Thin films consisting of blue-emitting quasi-2D perovskites generally show poor charge transport properties [40, 45].

The emissive layer comprises energetically coupled quantum-confined nanoparticles and quasi-2D phases, allowing excellent charge injection and suppressed non-radiative recombination required for efficient blue EL.

The above results indicate the coexistence of quasi-2D phases and isolated quantum-confined nanocrystals in the perovskite films.

The existence of excess PBA^+ cations in the perovskite films, presumably in the form of insulating PBABr, can impede charge injection into the perovskite films and influence the energetic coupling between the quasi-2D phases and the quantum-confined nanoparticles.

Restricting y to be in the range of 1.1 to 1.6 generates perovskite films with the structure of energetically coupled quantum-confined nanocrystals and quasi-2D phases.

The emissive layer consists of quantum-confined perovskite nanoparticles embedded in quasi-2D perovskites.

It is remarkable that a simple antisolvent processing scheme leads to efficient and reproducible EL from the quantum-confined all-bromide perovskite nanoparticles.

Methods

The TFB layers were deposited by spin-coating the polymer solution (in m-xylene, 8 mg ml^{-1}) at 2,000 r.p.m.

The perovskite films were prepared by spin-coating the precursor solution at 4,000 r.p.m.

The samples were characterized by a Cs aberration-corrected Titan G2 80–200 ChemiSTEM microscope operated at 200 kV. TEM observations on the quantum-sized nanoparticles were conducted using a Tecnai G2 F20 microscope operated at 200 kV. The TEM samples were prepared by dissolving the perovskite film (deposited onto substrates of ITO/PVK) in chlorobenzene [52], which dissolved the underlying PVK layers.

The X-ray radiation energy was 14 keV and the sample-to-detector distance was set to 2 m. A Pilatus 2M (Dectris Ltd) detector was used to collect GISAXS patterns that were measured at a grazing angle of ~0.1°.

The PL excitation spectra were collected by the FLS920 spectrometer with a xenon lamp as the source of excitation.

Modulation of Recombination Zone Position for Quasi-Two-Dimensional Blue Perovskite Light-Emitting Diodes with Efficiency Exceeding 5%

DOI: https://doi.org/10.1038/s41467-019-09011-5

This article is licensed under a Creative Commons Attribution 4.0 International License, which permits use, sharing, adaptation, distribution and reproduction in any medium or format, as long as you give appropriate credit to the original author(s) and the source, provide a link to the Creative Commons license, and indicate if changes were made. The images or other third party material in this article are included in the article's Creative Commons license, unless indicated otherwise in a credit line to the material. If material is not included in the article's Creative Commons license and your intended use is not permitted by statutory regulation or exceeds the permitted use, you will need to obtain permission directly from the copyright holder. To view a copy of this license, visit http://creativecommons.org/licenses/by/4.0/.

Originators: Li, Z., Chen, Z., Yang, Y. *et al.* Modulation of recombination zone position for quasi-two-dimensional blue perovskite light-emitting diodes with efficiency exceeding 5%. *Nat Commun* 10, 1027 (2019). https://doi.org/10.1038/s41467-019-09011-5

Abstract-Summary

Substantial progress has been made in developing perovskite light-emitting diodes with near-infrared, red and green emissions and over 20% external quantum efficiency.

The development of perovskite light-emitting diodes with blue emission remains a great challenge, which retards further development of full-color displays and white-light illumination based on perovskite emissive materials.

Through modulating the position of the recombination zone, we activate the majority of quasi-two-dimensional perovskite crystals, and thus demonstrate the most efficient blue perovskite light-emitting diode to date with emission peak at 480 nm, record luminance of 3780 cd m^{-2} and record external quantum efficiency of 5.7%.

Introduction

The blue photoluminescence (PL) and electroluminescence (EL) of perovskite thin films and LEDs can be achieved by forming quasi-two-dimensional (quasi-2D) perovskite from a three-dimensional (3D) bulk perovskite.

It is possible to achieve a more stable blue perovskite LED by forming a quasi-2D perovskite from a green-emissive all-inorganic 3D perovskite of $CsPbBr_3$.

The failure in achieving blue perovskite LED by forming quasi-2D perovskite based on a $CsPbBr_3$ system prompts us to enlarge the bandgap of the 3D component in quasi-2D perovskite films to obtain a blue EL regardless of the presence of highly efficient energy transfer and/or charge transfer from the 2D to 3D component.

On the basis of the recombination zone control in this hybrid layer, we activate most of the perovskite emitters and thus successfully demonstrate a highly efficient blue perovskite LED with maximum EQE of 5.7% and maximum luminance of 3780 cd m^{-2}, which, to our best knowledge, is a record performance for perovskite LEDs with blue emission.

Results

Although we dramatically increased the PEABr ratio from 0 to 120%, the XRD pattern from 3D perovskite was always present, also suggesting that the film we obtained was a quasi-2D perovskite film that combined with 2D and 3D perovskites.

The trend of EQE change with increasing PEABr ratios is similar to that of PLQYs of perovskite films, suggesting that the EL property of perovskite LEDs is governed by the PL property of the corresponding perovskite thin films.

A gradual blue shift of EL from 487 to 478 nm was observed as the PEABr ratio increased, which correlated well with the blue shift of PL spectra because of the gradual loss of the 3D nature of the perovskite film.

According to these previous studies, it can be expected that the position of the recombination zone for the injected charge carriers is critical to achieving a highly efficient perovskite LED if a vertically non-uniform distribution of nanostructure is existed in the perovskite thin film.

Discussion

We demonstrate a highly efficient blue perovskite LED by combining the strategies of composition engineering in quasi-2D perovskite crystals and modulation of the recombination zone position.

Our study shows that it is possible to further improve the performance of blue perovskite LEDs by carefully engineering the composition and dimension of the perovskite crystals, as well as the recombination zone position in the devices, paving the way for realization of full-color displays and white-light illumination with perovskite LEDs.

A pure-halide perovskite system with pure blue and red emission may be ultimately needed for stable perovskite LEDs.

Considering the ionic nature of perovskite crystals, the strong electric field in a thin film perovskite LED also could be another possible factor for device degradation, and this would be more severe for a blue perovskite LED as which needs higher applied voltage for operation.

Methods

After a 4-min oxygen plasma treatment for ITO, PEDOT:PSS (CH 8000) aqueous solution was spin-coated onto the ITO substrate at 3500 rpm for 50 s and baked at 150 °C for 15 min in ambient air.

Perovskite precursor solution was then spin-coated onto the PEDOT:PSS film at 2500 rpm for 2 min in an N_2-filled glovebox.

The crystalline structure of the perovskite films was investigated with an X-ray diffractometer (PANalytical X'pert PRO) equipped with a Cu-Kα X-ray tube, using ITO/PEDOT:PSS/perovskite as samples.

The SEM images were obtained with a Zeiss EVO 18 SEM based on the sample structure of ITO/PEDOT:PSS/perovskite.

The trap densities of the perovskite films were extracted by the dark current–voltage characteristics of the hole-only devices in the device architecture of the ITO/PEDOT:PSS/perovskite/MoO_3 (10 nm)/Ag (100 nm) through a computer-controlled Keithley 2400 source meter.

PLQYs of the perovskite films were recorded by a commercialized PLQY measurement system from Ocean Optics with excitation from a 365-nm LED.

Large Cation Ethylammonium Incorporated Perovskite for Efficient and Spectra Stable Blue Light-Emitting Diodes

DOI: https://doi.org/10.1038/s41467-020-17943-6

This article is licensed under a Creative Commons Attribution 4.0 International License, which permits use, sharing, adaptation, distribution and reproduction in any medium or format, as long as you give appropriate credit to the original author(s) and the source, provide a link to the Creative Commons license, and indicate if changes were made. The images or other third party material in this article are included in the article's Creative Commons license, unless indicated otherwise in a credit line to the material. If material is not included in the article's Creative Commons license and your intended use is not permitted by statutory regulation or exceeds the permitted use, you will need to obtain permission directly from the copyright holder. To view a copy of this license, visit http://creativecommons.org/licenses/by/4.0/.

Originators: Chu, Z., Zhao, Y., Ma, F. *et al.* Large cation ethylammonium incorporated perovskite for efficient and spectra stable blue light-emitting diodes. *Nat Commun* 11, 4165 (2020). https://doi.org/10.1038/s41467-020-17943-6

Abstract-Summary

Perovskite light-emitting diodes (PeLEDs) have showed significant progress in recent years; the external quantum efficiency (EQE) of electroluminescence in green and red regions has exceeded 20%, but the efficiency in blue lags far behind.

This method modulates the photoluminescence from the green region (508 nm) into blue (466 nm), and over 70% photoluminescence quantum yield in blue is obtained.

With configuration of PeLEDs with 60% EABr, as high as 12.1% EQE of sky-blue electroluminescence located at 488 nm has been demonstrated, which will pave the way for the full color display for the PeLEDs.

Introduction

Lead halide perovskite shows very high potential in display due to its unique emission properties, significant progresses in electroluminescence (EL) for perovskite light-emitting diodes (PeLEDs) have been achieved in recent years [1, 18–20, 27–57].

The effective attempts to obtain blue PeLEDs are generating the quantum-well structure via reducing-dimensional (quasi-2D and 0D) perovskites with different large cations [40, 42–59].

During preparation of this manuscript, Jin and others have reported efficient blue light-emitting diodes (LEDs) based on quantum-confined bromide perovskite via using phenylbutylammonium bromide as quasi-2D phases combined with an anti-solvent film deposition method, a peak EQE of up to 9.5% has been achieved [60].

The obtained perovskite films showed larger than 70% photoluminescence quantum yield (PLQY) in blue, and more importantly, the blue emission shows good spectra stability under light-soaking and heating.

By optimizing the content of EABr, a 12.1% EQE of the sky-blue (488 nm) EL was obtained, which could be the highest efficiency for blue emission PeLEDs so far.

Results

The quasi-2D $PEA_2(CsPbBr_3)_2PbBr_4$ perovskite films with different ratio of EABr were synthesized through mixing $PbBr_2$, CsBr, PEABr, and EABr in the solution of dimethyl sulfoxide (DMSO) and a one-step spin-coating process (details in "Methods").

The introduction of EABr lead to the decrease of the diffraction angles, indicating an expansion of the $CsPbBr_3$ crystal lattice by insertion of the larger EA cations.

Whiling introducing more EABr ($x \geq 0.75$) in $CsPbBr_3$, the films exhibit two photoluminescence peaks and an excitonic absorption, the higher energy emission and sharp absorption peaks is almost overlap with $EAPbBr_3$ itself, indicating that EA maybe not completely go into the $CsPbBr_3$ while the EA amount is too much.

The first-principle calculation of three-dimensional perovskite has confirmed the increasing of band gap with introduction of EA cation into $CsPbBr_3$ lattice.

Discussion

With the blue shift of EL spectra, the corresponding CIE value and the EL image of different EABr also changed from the green to blue region, indicating the successful achievement of a blue-emissive perovskite LED.

The 12.1% EQE for sky-blue emission at 488 nm could be the best for the PeLEDs.

We have demonstrated that the blue emission of perovskite LEDs with EA cation incorporation into three-dimensional perovskite lattice.

The emission has been tuned from green into blue, and showed high PLQY (>70%).

With tailoring the EA composition, an efficient PeLEDs with 12.1% EQE in 488 nm emission has been achieved.

Methods

The substrates were transferred into a nitrogen-filled glove box, and the perovskite precursor solution was then spin-coated on the m-PEDOT:PSS film at 3000 rpm for 2 min, followed by annealing on a hot plate at 60 °C for 20 min.

Steady-state PL spectra of the perovskite films was measured at room temperature in the ambient air using a FLS1000 spectrometer.

The work function Φ (that is the Fermi level absolute value of the free film) of the test films can be calculated from following equation: $h\nu - \Phi = E_{Fermi} - E_{cutoff}$, where E_{Fermi} and E_{cutoff} is the value of Fermi level position and the steep edge position in UPS spectrum, respectively, $h\nu = 21.22$ eV and $E_{Fermi} = 21.51$ eV. The I-V-L curve, EL spectrum, EQE, CIE, and operating lifetime of the perovskite LED were carried out simultaneously by a commercialized system (LQE-50-EL, Enlitech) that was equipped with an integrated sphere and photomultiplier tubes (PTM), in which the PMT is used to measure the low luminance.

Smoothing the Energy Transfer Pathway In Quasi-2D Perovskite Films Using Methanesulfonate Leads to Highly Efficient Light-Emitting Devices

DOI: https://doi.org/10.1038/s41467-021-21522-8

This article is licensed under a Creative Commons Attribution 4.0 International License, which permits use, sharing, adaptation, distribution and reproduction in any medium or format, as long as you give appropriate credit to the original author(s) and the source, provide a link to the Creative Commons license, and indicate if changes were made. The images or other third party material in this article are included in the article's Creative Commons license, unless indicated otherwise in a credit line to the material. If material is not included in the article's Creative Commons license and your intended use is not permitted by statutory regulation or exceeds the permitted use, you will need to obtain permission directly from the copyright holder. To view a copy of this license, visit http://creativecommons.org/licenses/by/4.0/.

Originators: Kong, L., Zhang, X., Li, Y. *et al.* Smoothing the energy transfer pathway in quasi-2D perovskite films using methanesulfonate leads to highly efficient light-emitting devices. *Nat Commun* 12, 1246 (2021). https://doi.org/10.1038/s41467-021-21522-8

Abstract-Summary

We make use of methanesulfonate (MeS) that can interact with the spacer BA cations via strong hydrogen bonding interaction to reconstruct the quasi-2D perovskite structure, which increases the energy acceptor-to-donor ratio and enhances the energy transfer in perovskite films, thus improving the light emission efficiency.

Green light-emitting diodes fabricated using these quasi-2D RP perovskite films reach current efficiency of 63 cd A^{-1} and 20.5% external quantum efficiency, which are the best reported performance for devices based on quasi-2D perovskites so far.

Introduction

Quasi-two-dimensional (quasi-2D) perovskites offer larger exciton binding energy (hundreds of meV) and lower ion mobility compared with 3D perovskites [61–67], making them even more promising light-emitting materials.

There are two key points that determine the emission efficiency of quasi-2D RP perovskites, namely the exciton energy transfer process and the deep defect state formation derived from the bandgap broadening [65–67].

The distribution of energy domains is inhomogeneous in quasi-2D perovskites due to the random stacking of low-dimensional structures with varying degrees of quantum and dielectric confinement [68–70].

We seek to realize the full potential of quasi-2D perovskites for LEDs by smoothing the energy transfer pathway through structure reconstruction and defect reduction using methanesulfonate (MeS) additive.

Quasi-2D green PeLEDs fabricated following this strategy show current efficiency (CE) of 63 cd A^{-1} and up to 20.5% external quantum efficiency (EQE), making this device the best-performing quasi-2D perovskite LED reported in literature so far.

Results

The introduction of MeS in precursor solution can control the crystallization growth kinetics of quasi-2D perovskite films resulting in enhanced exciton energy transfer from small-n phases to large-n phases, which will be discussed below in detail.

To address the influence of MeS on the perovskite phase distribution, measurements of the X-ray photoelectron spectra (XPS) that track the N signal (where BA is the only source of that atomic species and thus the signal stands for the phase distribution changes) along with decreasing the perovskite film thickness by increasing the Ar^+ etching time were conducted.

The decay kinetics for $GSB_{n\,\geq 4}$ peak of 8%-MeS-treated perovskite film shows a shorter formation time of 910 fs, demonstrating that an accelerated exciton energy transfer occurs inside the reconstructed phases enabling more efficient energy transfer from small-n phases to large-n phases.

Discussion

We achieved a record efficiency (CE of 63 cd A^{-1} and EQE of 20.5%) for quasi-2D $BA_2Cs_{n-1}Pb_nBr_{3n+1}$ PeLEDs by introducing MeS into the perovskite precursor solutions to reconstruct the perovskite phase distribution and reduce uncoordinated Pb^{2+} ions, and thus enhance the emission efficiency of perovskite films derived from these solutions.

The proportion of large-n phases with narrow bandgap has increased in the distribution phases of different n values of quasi-2D perovskites due to hydrogen-bonding interaction between $-SO_3^-$ group of MeS and BA spacer; and the surface uncoordinated Pb^{2+} ions on the perovskites were simultaneously reduced benefiting from the electron-rich Lewis base nature of MeS. Besides, the blocking effect of BA which tends to induce defects during crystal growth could be reduced because of the interaction between BA and MeS. This work offers a promising approach for fabricating next-generation high-performance PeLEDs based on quasi-2D perovskites.

Methods

TFB, TPBI, and LiF were purchased from Luminescence Technology Corp. Absolute ethyl alcohol and toluene were purchased from Sinopharm Chemical Reagent Co., Ltd. $PbBr_2$, CsBr, and BABr were dissolved in anhydrous DMSO with the ratio of 1:1.2:0.4, and PEO (3.33 mg ml^{-1}) was added to improve the morphology of the resulting film [71–73].

A TFB:PVK layer was spin-coated onto oxygen-plasma-treated ITO substrates at 4000 rpm for 40 s, followed by annealing at 150 °C for 20 min.

PVP (1.5 mg ml^{-1} in absolute ethyl alcohol) was spin-coated on top of the TFB:PVK layer.

^{1}H NMR spectra were recorded on a Bruker Advanced II (400 MHz) spectrometer at room temperature.

The calculated ground-state energies are −90.935, −46.738, −133.728, and −6.107 eV for BABr, CsMeS, BAMeS, and CsBr, respectively.

High-Performance Large-Area Quasi-2D Perovskite Light-Emitting Diodes

DOI: https://doi.org/10.1038/s41467-021-22529-x

This article is licensed under a Creative Commons Attribution 4.0 International License, which permits use, sharing, adaptation, distribution and reproduction in any medium or format, as long as you give appropriate credit to the original author(s) and the source, provide a link to the Creative Commons license, and indicate if changes were made. The images or other third party material in this article are included in the article's Creative Commons license, unless indicated otherwise in a credit line to the material. If material is not included in the article's Creative Commons license and your intended use is not permitted by statutory regulation or exceeds the permitted use, you will need to obtain permission directly from the copyright holder. To view a copy of this license, visit http://creativecommons.org/licenses/by/4.0/.

Originators: Sun, C., Jiang, Y., Cui, M. *et al.* High-performance large-area quasi-2D perovskite light-emitting diodes. *Nat Commun* 12, 2207 (2021). https://doi.org/10.1038/s41467-021-22529-x

Abstract-Summary

We investigate the failure mechanism of the widespread active film fabrication method; and ascribe severe phase-segregation to be the reason.

High-quality large-area quasi-2D films with desirable properties are obtained.

We reported a series of highly-efficient green quasi-2D PeLEDs with active areas of 9.0 cm^2.

The peak EQE of 16.4% is achieved in = 3, represent the most efficient large-area PeLEDs yet.

High brightness device with luminance up to 9.1×10^4 cd m^{-2} has achieved in = 10 film.

Introduction

Quasi-2D perovskite films with self-assembled multiple-quantum-well structure represent an important category of perovskites and have achieved great success in small-area PeLEDs due to their outstanding optical properties [18, 19, 57, 79–81].

Many factors are still unclear in large-area quasi-2D PeLEDs fabrication.

Verifying the validity of the "antisolvent-assisted" approach, exploring the failure mechanism, and seeking new solutions are essential steps for developing high-performance, large-area quasi-2D PeLEDs.

We conclude that the key to generate efficient large-area quasi-2D PeLEDs is to suppress the severe phase-segregation, particularly in the edge-region of the device.

The new intermediate phase does change the crystallization pathway and facilitate the formation of large-area, high-quality quasi-2D films with desirable optical and electrical properties.

Taking advantage of the above findings, here we reported a series of highly efficient, large-area quasi-2D PeLEDs with active areas of 9.0 cm^2.

Results

The "antisolvent-assisted" approach was employed to fabricate large-area = 5 quasi-2D, $PEA_2(FA_{0.7}Cs_{0.3})_4Pb_5Br_{16}$, perovskite films to verify its validity in large-area device fabrication.

The film fabricated with NVAL additive displays characteristic quasi-2D GSB peaks, but in the absence of n = 1 and 3D phases.

By adjusting -values, the corresponding quasi-2D perovskite films' emitting behavior can be regulated.

To further understand the relationship between material's E_b and corresponding carrier recombination kinetics, we fabricated quasi-2D films, $PEA_2(FA_{0.7}Cs_{0.3})_{n-1}Pb_nBr_{3n+1}$, with = 3, 5, 7, and 10.

The low -value films cannot reserve high PLQY under high current density because of the effective Auger recombination, thus limiting their application in high-brightness required devices, such as solid-state lighting.

High -value films possess improved conductivity for better injection and transport; more importantly, lower PLQY roll-off was observed thanks to the reduced Auger recombination rates.

Discussion

We have developed an "antisolvent-free" method to fabricate large-area quasi-2D perovskite films, which is achieved by partially anchoring amphiphilic NVAL molecules on $[PbBr_6]^{4-}$ inorganic slabs.

The approach overcomes the phase-segregation, enabling high-quality large-area quasi-2D films becomes possible.

Taking advantage of the above findings, we achieved a series of highly efficient, large-area quasi-2D PeLEDs with active areas of 9.0 cm^2.

The work demonstrates quasi-2D perovskites' great potential and diversity for display or lighting application, thus it paved the way for future large-area PeLEDs manufacture.

Methods

PFNBr dissolved in methanol with the concentration of 0.5 mg/mL was then spin-coated on top of the PEDOT:PSS layer for 1 min at 4000 r.p.m.

The PLQY were measured using an integrating sphere method, a 365-nm laser was used to photo-excite the films, and a Quanta-Phi integrating sphere with a Fluorolog system was used to collect the spectra.

The TRPL spectra were measured by using a spectrometer (FLS980) with a picosecond pulsed laser under a repetition rate of 800 kHz at 355 nm.

^{1}H NMR spectra were recorded in DMSO-d6 solvent by using a Bruker 400 MHz NMR spectrometer.

The UPS spectra were recorded using an Omicron ultrahigh-vacuum chamber with excitation provided by the He I emission line (21.2 eV) of a helium pressure about 2×10^{-8} mbar.

High-Performance Quasi-2D Perovskite Light-Emitting Diodes: From Materials to Devices

DOI: https://doi.org/10.1038/s41377-021-00501-0

This article is licensed under a Creative Commons Attribution 4.0 International License, which permits use, sharing, adaptation, distribution and reproduction in any medium or format, as long as you give appropriate credit to the original author(s) and the source, provide a link to the Creative Commons license, and indicate if changes were made. The images or other third party material in this article are included in the article's Creative Commons license, unless indicated otherwise in a credit line to the material. If material is not included in the article's Creative Commons license and your intended use is not permitted by statutory regulation or exceeds the permitted use, you will need to obtain permission directly from the copyright holder. To view a copy of this license, visit http://creativecommons.org/licenses/by/4.0/.

Originators: Zhang, L., Sun, C., He, T. *et al.* High-performance quasi-2D perovskite light-emitting diodes: from materials to devices. *Light Sci Appl* 10, 61 (2021). https://doi.org/10.1038/s41377-021-00501-0

Abstract-Summary

Quasi-two-dimensional (quasi-2D) perovskites have attracted extraordinary attention due to their superior semiconducting properties and have emerged as one of the most promising materials for next-generation light-emitting diodes (LEDs).

The review herein presents an overview of the inherent properties of quasi-2D perovskite materials, the corresponding energy transfer and spectral tunability methodologies for thin films, as well as their application in high-performance LEDs.

The review thus provides a systematic and timely summary for the community to deepen the understanding of quasi-2D perovskite materials and resulting LED devices.

Introduction

Three types of perovskite materials with different dimensions (i.e., 3D perovskites, quasi-2D perovskites, and perovskite nanocrystals) are commonly included in the emitter layer of PeLEDs [50–51, 82, 83, 84].

Quasi-2D perovskites represent an important category of perovskites that possess self-assembled multi-quantum-well structures and have gained great success in light emission applications owing to their outstanding optoelectrical properties [62, 85].

Substantial efforts have been made to obtain high-performance quasi-2D PeLEDs, which have facilitated unprecedented rapid development.

Five years, we have witnessed the rapid development of quasi-2D perovskite optoelectronics, especially their tremendous success in LED applications.

Quasi-2D perovskites exhibit unique optical properties arising from their structural characteristics, which are different from those of conventional 3D [1, 31–87] and two-dimensional (2D) perovskites [88, 89].

We then summarize the progress in spectral tunability of quasi-2D perovskites, mainly to realize high-performance pure-red and pure-blue emission.

We discuss the newly emerged device engineering approaches to produce high-performance quasi-2D PeLEDs.

Characteristics of quasi-2D perovskites

Quasi-2D perovskites possess the chemical formula $A'_2A_{n-1}B_nX_{3n+1}$ ($1 \leq n \leq \infty$), where A′ refers to a large organic cation, including monoammonium cations (R-NH_3^+) and diammonium cations (^+H_3N-R-NH_3^+) (R represents an alkyl chain or aromatic ligand); A stands for a small monovalent cation, e.g., methylammonium ($MA^+ = CH_3NH_3^+$), formamidine ($FA^+ = CH(NH_2)_2^+$), or cesium (Cs^+); B is a divalent metal cation such as lead (Pb^{2+}) or tin (Sn^{2+}); X represents a halide, e.g., chloride (Cl^-), bromide (Br^-) or iodide (I^-); and n refers to the number of $[BX_6]^{4-}$ octahedral units.

Due to the quantum- and dielectric-confinement effects arising from the QW structure, the E_b of a quasi-2D perovskite becomes larger than that of its 3D analog [19, 62].

The carrier recombination dynamics of quasi-2D perovskite single crystals with various n-values were systematically studied to deeply understand the photophysical properties of quasi-2D perovskites.

They found that the existence of excitons and free carriers varied in quasi-2D perovskite single crystals with different n-values.

In low n-value quasi-2D perovskites, E_b is large, which guarantees efficient exciton recombination.

Recombination characteristics of quasi-2D perovskite films

The energy transfer facilitates radiative recombination, resulting in high PLQYs for quasi-2D perovskite films even at low pumping densities.

From the perspective of the crystallization kinetics and the carrier recombination characteristics of quasi-2D perovskite films, we conclude that efficient energy transfer, effective exciton recombination, and low defect density are the most striking features for quasi-2D perovskite films, which contribute to the excellent optical properties [90].

For quasi-2D perovskite films, the n-value distribution affects their energy transfer and corresponding radiative exciton recombination efficiency.

To the n-value distribution and the grain orientations, the coupled quasi-2D perovskite phases can also affect the energy transfer.

Enhancing radiative exciton recombination by increasing E_b can also improve the optical properties of quasi-2D perovskite films.

Since the crystallization of quasi-2D perovskite films is a self-assembly process, the precursor composition strongly influences the surface state, grain boundaries, and phase distribution of the ultimate films.

Color-pure emission of quasi-2D perovskites

Increasing the content of large organic cations can monotonically reduce the value of quasi-2D perovskite films.

This does not mean that quasi-2D perovskite films with low values are sufficient to effectively realize pure red or blue emission.

The resultant films displayed high PLQYs and stable blue emission by modulating the phase judiciously, thereby fabricating efficient and spectrally stable sky-blue quasi-2D PeLEDs.

After a series of artificial designs and interventions regarding the phase distribution, the optical properties of pure red and blue quasi-2D perovskite films have significantly improved, paving the way for high-performance pure red and blue quasi-2D PeLED manufacture.

Successful fabrication of high-performance quasi-2D PeLEDs with pure red and blue emission that satisfy display purposes might require a combination of strategies leveraging anion engineering, cation engineering of the "A-site" or "B-site," and dimensionality engineering.

High-performance quasi-2D PeLEDs

Device engineering, such as optimization of the charge transport layers of quasi-2D PeLEDs, could promote the charge injection balance factor towards its maximum ($\gamma = 1$) [91].

To convert high PLQYs of quasi-2D perovskite films into high EQEs of quasi-2D PeLEDs, the band alignment of the device structure is the most basic consideration.

We summarize the effects of organic cation spacers and grain orientations on the carrier transport and recombination capabilities of quasi-2D perovskite films when the device is operated under a bias voltage.

The random grain orientations in the quasi-2D perovskite films forced contact between the inorganic layers to improve charge transport and radiative recombination [92].

The LiF layer acted as a useful template for the growth of high-quality perovskite films and enhanced the device performance for green quasi-2D/3D PeLEDs.

Increasing the interaction between large organic cations and inorganic layers is a feasible strategy to improve the operational stability of quasi-2D PeLEDs.

Conclusions and perspectives

The advances in quasi-2D PeLEDs indicate their bright future in the application scenarios of ultrahigh-definition displays, solid-state lighting, optical communications, etc However, despite these impressive achievements, quasi-2D perovskite materials still have many unsettled issues that hinder their further development and application.

We discuss the prospects for the future development of novel quasi-2D perovskite materials/structures and white-light-emitting quasi-2D devices and the potential applications of quasi-2D perovskite emitters in large-area, printable, and flexible electronics as well as quasi-2D perovskite lasers, aiming to shed light on these promising future prospects.

The quasi-2D perovskite films with high PLQY and nucleation density show unique potential for fabricating large-area LED devices.

The performance of large-area (12.8 cm^2) quasi-2D PeLEDs is still far below that of small-area devices reported [93].

Quasi-2D perovskite films are compatible with low-temperature solution-based manufacturing techniques, such as inkjet, roll-to-roll, and 3D printing, also providing great potential in large-area and flexible electronics.

Advances and Challenges in Two-Dimensional Organic–Inorganic Hybrid Perovskites Toward High-Performance Light-Emitting Diodes

DOI: https://doi.org/10.1007/s40820-021-00685-5

This article is licensed under a Creative Commons Attribution 4.0 International License, which permits use, sharing, adaptation, distribution and reproduction in any medium or format, as long as you give appropriate credit to the original author(s) and the source, provide a link to the Creative Commons license, and indicate if changes were made. The images or other third party material in this article are included in the article's Creative Commons license, unless indicated otherwise in a credit line to the material. If material is not included in the article's Creative Commons license and your intended use is not permitted by statutory regulation or exceeds the permitted use, you will need to obtain permission directly from the copyright holder. To view a copy of this license, visit http://creativecommons.org/licenses/by/4.0/.

Originators: Ren, M., Cao, S., Zhao, J. *et al.* Advances and Challenges in Two-Dimensional Organic–Inorganic Hybrid Perovskites Toward High-Performance Light-Emitting Diodes. *Nano-Micro Lett.* 13, 163 (2021). https://doi.org/10.1007/s40820-021-00685-5

Abstract-Summary

The advantages and challenges of 2D perovskite light-emitting diodes (PeLED) have been also discussed, which may benefit the audient to get insight into the 2D perovskite materials as well as the resultant LED devices.

Introduction

2D perovskite materials have unique physical and chemical properties due to the influences of quantum confinement and dielectric confinement effect [94].

2D perovskites have become star materials used in many fields, including LEDs [95, 96], solar cells [97–100], photodetectors [101, 102], lasers [103, 104] and sensors [105].

This makes quasi-2D perovskite materials promising for high-performance LEDs.

Lee and others [106] prepared high-quality quasi-2D perovskite films by a simple and novel encapsulation growth method and obtained high-efficiency sky-blue LEDs with maximum EQE of 12.8%.

Di and others [50] reported near-infrared LEDs with quasi-2D and 3D perovskites and insulating polymers as emissive layers.

Su and others [107] proposed a co-interlayer engineering strategy for the preparation of quasi-2D perovskite materials, which makes the green LEDs have a high current efficiency of 66.1 cd A^{-1}.

The unique optical properties of 2D perovskite are discussed as the advantages of the emitting layer of LEDs.

Characteristics of 2D Perovskite Materials

The quasi-2D perovskites are a mixture of many phases (especially a large number of small n phases), resulting in inefficient energy transfer.

Although the mechanism of photogenerated electron and hole transfer is still controversial, the rapid carrier transfer process can compete with trap-assisted monomolecular non-radiative recombination, which makes the quasi-2D perovskite structure beneficial to enhance PLQY [108, 109].

3D perovskite thin films have rough surface and are prone to a large number of pinholes, leading to the increase of leakage current and obvious thermal effect, which will be resulting in the decrease of device efficiency and the shortening of device life [31, 110].

Zeng and others [111] used organic molecules to passivate the top and bottom of perovskite quantum dot film, which improved the stability of PeLEDs and effectively suppressed the non-radiative recombination.

In quasi-2D perovskite, Auger recombination can be suppressed by reducing the local carrier density in the quantum well.

Strategies for Boosting the Performance of PeLEDs

The quasi-2D perovskite materials suffer from serious efficiency roll-off, which is mainly caused by the charge injection imbalance, Auger recombination and Joule heating [112].

Although balanced charge injection can suppress the efficiency roll-off, there is still a serious efficiency roll-off at a high current density, which may be due to the existence of long organic chains, resulting in poor charge transport, or the increase of Auger recombination rate due to the quantum well structures.

The chain lengths and $\pi - \pi$ stacking of organic cations affect the n distribution of 2D perovskites [45–114].

Through comparison, it was found that different organic amine cations would affect the thicknesses of perovskite quantum wells, efficiency of energy transfer and recombination kinetics, resulting in great differences in the performance of PeLEDs [115].

As we known, effective energy transfer plays a decisive role in the photoluminescence efficiency of 2D perovskites.

Current Challenges Ahead in 2D PeLEDs

It is also an important strategy to achieve blue emission by inserting large organic cations into 3D perovskites.

Ion migration is the main factor leading to poor stability of 2D PeLEDs.

Enhancing the interaction between large organic cations and octahedral slabs is beneficial to improve the stability of 2D PeLEDs.

Most of the researches on 2D perovskite optoelectronic devices are mainly based on lead-based perovskites at the moment.

Dou and others used 7-(thiophen-2-yl)benzothiadiazol-4-yl)-[2,2'bithiophen]-5-yl)ethylammonium iodide (BTmI) as the barrier layers to display a pure red Sn-based PeLED with a luminance of 3466 cd m^{-2} and a working stability of more than 150 h [116]. (111)-oriented 2D perovskites have also been used as LEDs, but they all show unsatisfactory efficiency.

Even though the performances of lead-free PeLEDs are far from that of lead-based perovskite, there is a great room for development.

Conclusions and Outlook

The structure of 2D perovskites can be regarded as the periodic splitting of 3D perovskites along (100), (110) and (111) planes by organic cations.

We analyze and compare the structures of 2D perovskites with different orientations.

2D perovskites have many excellent properties, such as convenient control of the width of quantum wells, formation of dense films and ultra-fast energy transfer, which shows great potential in the field of optoelectronic devices.

Although many alkylamines or aromatic amines have been explored as spacers, there is still a lot of space to expand the 2D perovskites family, which is helpful to understand the 2D perovskites comprehensively and to grasp the influence of perovskites structures and properties on device performance fundamentally.

In 2D perovskite materials, although large organic cations can improve the humidity resistance of films, the stability of devices is still threatened by ion migration, thermal instability and interface instability.

Suppressing Thermal Quenching via Defect Passivation for Efficient Quasi-2D Perovskite Light-Emitting Diodes

DOI: https://doi.org/10.1038/s41377-022-00761-4

This article is licensed under a Creative Commons Attribution 4.0 International License, which permits use, sharing, adaptation, distribution and reproduction in any medium or format, as long as you give appropriate credit to the original author(s) and the source, provide a link to the Creative Commons license, and indicate if changes were made. The images or other third party material in this article are included in the article's Creative Commons license, unless indicated otherwise in a credit line to the material. If material is not included in the article's Creative Commons license and your intended use is not permitted by statutory regulation or exceeds the permitted use, you will need to obtain permission directly from the copyright holder. To view a copy of this license, visit http://creativecommons.org/licenses/by/4.0/.

Originators: Zhang, D., Fu, Y., Zhan, H. *et al.* Suppressing thermal quenching via defect passivation for efficient quasi-2D perovskite light-emitting diodes. *Light Sci Appl* 11, 69 (2022). https://doi.org/10.1038/s41377-022-00761-4

Abstract-Summary

Emission thermal quenching is commonly observed in quasi-2D perovskite emitters, which causes the severe drop in luminescence efficiency for the quasi-2D perovskite light-emitting diodes (PeLEDs) during practical operations.

This issue is often neglected and rarely studied, and the root cause of the thermal quenching has not been completely revealed now.

The agent can effectively passivate coordination-unsaturated Pb^{2+} defects of both surface and bulk of the film without affecting the perovskite crystallization, which helps to more truly demonstrate the important role of defects in thermal quenching.

Our results reveal the root cause that the quenching will be strengthened by the defect-promoted exciton-phonon coupling.

Introduction

The quasi-2D perovskites as the emitters usually suffer from a problem of thermal quenching that is easy to be ignored.

The photoluminescence (PL) thermal quenching has been found in 3D, quasi-2D, and nanocrystals (NCs) perovskites [117–121].

As for the quasi-2D perovskites, the thermal quenching and its inhibition have not been systematically reported yet.

Considering their unique luminescence mechanism with energy funnel, the effect of defects in emitting region and energy transfer region of the quasi-2D perovskite film on device performance and thermal stability should be explored.

To reveal the direct correlation between thermal quenching and defects, we develop a phosphate fluorene passivation agent with functional groups that can effectively passivate the coordination-unsaturated Pb^{2+} defect, while the crystallization of quasi-2D perovskite is not affected at all after passivation.

We reveal that the defects will induce the thermal quenching by strengthening the exciton-phonon interaction in quasi-2D perovskite.

Results

Based on nonpolar fluorene and polar phosphate groups, the DBPF can be dissolved into perovskite precursor solution (dimethyl sulfoxide, DMSO) and antisolvent (ethyl acetate, EA) to achieve bulk passivation and surface passivation, which is benefit to further explore the relationship between defects in different landscapes and device performance including thermal stability.

The bulk and surface passivation for the perovskite film are based on the same mechanism of coordinating with Pb^{2+}, but their impacts on device performance and thermal quenching may not be consistent.

These evidences can fully verify that the nonradiative recombination is closely related to the defects in quasi-2D perovskites, and defect passivation for the emitting region of the film surface is the key to weaken the nonradiative recombination and improve the PLQY.

Our passivation strategy improves the performance of the device working at high temperature, and it is a direct evidence to prove that the thermal quenching is closely related to the defects.

Discussion

We have revealed that defects can cause stronger exciton-phonon coupling, which induces the thermal quenching in quasi-2D perovskite.

Research results demonstrate the correlation between passivation for different landscapes and device performances, as well as the thermal quenching.

Over 22% of EQE and doubled operation lifetime are obtained for PeLEDs based on dual passivation at RT, and the PeLEDs still maintain 85% of the initial EQE at high temperature of 85 °C.

Materials and methods

A solution of 2,7-dibromo-9,9-bis(3′-bromopropyl)-fluorene (2.83 g, 5 mmol) in triethyl phosphite was heated to 140 °C for 18 h in nitrogen atmosphere.

The residue was purified by eutral Al_2O_3 using ethyl acetate/petroleum ether (1:1) as eluent to obtain white granular microcrystals (2.8 g, 82%).

All quasi-2D perovskite layers are obtained by spin-coating.

The 0.2 M (Pb^{2+} concentration) precursor solution is spin-coated at 8000 rpm for 30 s, and then 100 μL of ethyl acetate is poured onto the film during the spin-casting, followed by annealing on a hot plate at 85 °C for 20 min.

PVK solution (10 mg ml^{-1} in chlorobenzene) is spin-coated at 4000 rpm for 40 s and the films are baked at 120 °C for 30 min.

The perovskite film is deposited on PVK layer.

UV-Vis absorption spectra of quasi-2D perovskite layers were obtained by a Perkin-Elmer Lambda 35 UV-vis spectrometer.

Suppressing Phase Disproportionation in Quasi-2D Perovskite Light-Emitting Diodes

DOI: https://doi.org/10.1038/s41467-023-36118-7

This article is licensed under a Creative Commons Attribution 4.0 International License, which permits use, sharing, adaptation, distribution and reproduction in any medium or format, as long as you give appropriate credit to the original author(s) and the source, provide a link to the Creative Commons license, and indicate if changes were made. The images or other third party material in this article are included in the article's Creative Commons license, unless indicated otherwise in a credit line to the material. If material is not included in the article's Creative Commons license and your intended use is not permitted by statutory regulation or exceeds the permitted use, you will need to obtain permission directly from the copyright holder. To view a copy of this license, visit http://creativecommons.org/licenses/by/4.0/.

Originators: Wang, K., Lin, ZY., Zhang, Z. *et al.* Suppressing phase disproportionation in quasi-2D perovskite light-emitting diodes. *Nat Commun* 14, 397 (2023). https://doi.org/10.1038/s41467-023-36118-7

Abstract-Summary

Electroluminescence efficiencies and stabilities of quasi-two-dimensional halide perovskites are restricted by the formation of multiple-quantum-well structures with broad and uncontrollable phase distributions.

We demonstrate that extending the π-conjugation length and increasing the cross-sectional area of the ligand enables perovskite thin films with dramatically suppressed ion transport, narrowed phase distributions, reduced defect densities, and enhanced radiative recombination efficiencies.

Our devices also exhibit wide wavelength tunability and improved spectral and phase stability compared with existing perovskite light-emitting diodes.

Introduction

The efficiency and stability are likely limited by the low phase purity and high defect densities of typical 2D perovskite films.

Such a broad n-distribution in quasi-2D perovskites could be understood as a result from phase "disproportionation" of an initial n-value material to yield lower and higher n-phases, which leads to phase broadening and instability over time.

Films with mixed phases exhibit cascade energy transfer from low-n to high-n phases [18, 19], resulting in emission that is dominated by the composition of higher-n phase and ultimately difficult to control.

A few recent efforts have been devoted to regulating phase distribution and smoothing the energy landscape in quasi-2D perovskite thin films [45, 107–125], but has exhibited limited success in wavelength-tunable LEDs due to ineffective phase control.

Results

The free energies required for PPT' and PPT are twice those of BA and TEA, indicating that these bulky organic conjugated ligands can effectively inhibit direct interlayer ion diffusion and potentially also hinder diffusion limited phase disproportionation.

To assess the impact of these ligands on phase distribution, they were used to fabricate lead iodide-based quasi-2D perovskite ($n > 1$) thin films that were subjected to multi-mode characterizations.

The PPT'-based film exhibits distinct absorption peaks at 572, 633, and 682 nm corresponding to $n = 2$, 4, and 5 phases, respectively.

The faster energy transfer processes may be associated with a combination of fewer transfer steps (better phase purity) and reduced carrier-trapping (lower defect density) in the PPT film [50].

The reduced defect density along with improved PLQY make PPT-based quasi-2D perovskite thin films promising for efficient LEDs.

Discussion

We have demonstrated that suppressed phase disproportionation leads to highly efficient and wavelength-tunable quasi-2D perovskite LEDs with a peak EQE of 26.3% and good stability and reproducibility.

This is the highest efficiency reported so far for a quasi-2D halide perovskite LED in the red region.

We have found that ionic transport is an important step of solid-phase disproportionation reaction in quasi-2D perovskites, and the key functions of those bulky organic conjugated ligands are to stabilize the perovskite kinetic products and inhibit ion diffusion.

Methods

The quasi-2D perovskite films were prepared by spin-coating the precursor solutions onto the above substrates at 4000 rpm for 60 s and annealing at 100 °C for 10 min.

Quasi-2D perovskite thin films were spin-coated on pre-cleaned glass substrates for XRD measurements.

The spectra were collected with a H Lyman-α photon source (E-LUX 121, photon energy of 10.2 eV) and a 5.85 eV pass energy while the samples were biased at negative 5 V. The optical properties of Poly-TPD, PPT based quasi-2D perovskite, TPBi, and Al were obtained using a J. A. Woollam V-Vase UV-vis-NIR spectroscopic ellipsometer from 500 to 800 nm.

The depth calibrations have been established using the interface tool in SurfaceLab version 7.2 software from ION-TOF GmbH to identify the different interfaces and based on the measured thicknesses using the surface profiler to obtain a line scan of the craters with the in-situ SPM by contact scanning.

Bibliography

1. Tan, Z.-K., et al.: Bright light-emitting diodes based on organometal halide perovskite. Nat. Nanotechnol. **9**, 687–692 (2014)
2. Yamada, Y., et al.: Photocarrier recombination dynamics in perovskite $CH_3NH_3PbI_3$ for solar cell applications. J. Am. Chem. Soc. **136**, 11610–11613 (2014)
3. Ishihara, T., Takahashi, J., Goto, T.: Optical properties due to electronic transitions in two-dimensional semiconductors $(CnH_{2n+1}NH_3)PbI_4$. Phys. Rev. B **42**, 11099–11107 (1990)
4. Mitzi, D.B.: In: Karlin, K.D. (ed.) Progress in Inorganic Chemistry, vol. 48, pp. 1–121. Wiley (1999)
5. Quan, L.N., Yuan, M., Comin, R., Voznyy, O., Beauregard, E.M., Hoogland, S., Buin, A., Kirmani, A.R., Zhao, K., Amassian, A., Kim, D.H., Sargent, E.H.: Ligand-stabilized reduced-dimensionality perovskites. J. Am. Chem. Soc. **138**(8), 2649–2655 (2016). https://doi.org/10.1021/jacs.5b11740
6. Stranks, S.D., et al.: Recombination kinetics in organic-inorganic perovskites: excitons, free charge, and subgap states. Phys. Rev. Appl. **2**, 034007 (2014)
7. Dou, L., et al.: Atomically thin two-dimensional organic-inorganic hybrid perovskites. Science **349**, 1518–1521 (2015)
8. Jeon, N.J., et al.: Solvent engineering for high-performance inorganic–organic hybrid perovskite solar cells. Nat. Mater. **13**, 897–903 (2014)
9. Xu, F., et al.: Efficient exciton funneling in cascaded PbS quantum dot supperstructures. ACS Nano **12**, 9950–9957 (2011)
10. Sichert, J.A., Tong, Y., Mutz, N., Vollmer, M., Fischer, S., Milowska, K.Z., García Cortadella, R., Nickel, B., Cardenas-Daw, C., Stolarczyk, J.K.: Quantum size effect in organometal halide perovskite nanoplatelets. Nano Lett. **15**(10), 6521–6527 (2015). https://doi.org/10.1021/acs.nanolett.5b02985
11. Supran, G.J., et al.: High-performance shortwave-infrared light-emitting devices using core–shell (PbS–CdS) colloidal quantum dots. Adv. Mater. **27**, 1437–1442 (2015)
12. Sun, L., et al.: Bright infrared quantum-dot light-emitting diodes through inter-dot spacing control. Nat. Nanotech. **7**, 369–373 (2012)

13. Hong, X., Ishihara, T., Nurmikko, A.: Dielectric confinement effect on excitons in PbI_4-based layered semiconductors. Phys. Rev. B **45**(12), 6961 (1992). https://doi.org/10.1103/PhysRevB.45.6961
14. Tanaka, K., Kondo, T.: Bandgap and exciton binding energies in lead-iodide-based natural quantum-well crystals. Sci. Technol. Adv. Mater. **4**, 599–604 (2003)
15. Smith, I.C., Hoke, E.T., Solis-Ibarra, D., McGehee, M.D., Karunadasa, H.I.: A layered hybrid perovskite solar-cell absorber with enhanced moisture stability. Angew. Chem. Int. Ed. **53**(42), 11232–11235 (2014)
16. de Mello, J.C., Wittmannn, H.F., Friend, R.H.: An improved experimental determination of external photoluminescence quantum efficiency. Adv. Mater. **9**, 230 (1997)
17. Era, M., et al.: Electroluminescent device using two dimensional semiconductor $(C_6H_5C_2H_4NH_3)_2PbI_4$ as an emitter. Synth. Met. **71**, 2013–2014 (1995)
18. Yuan, M., Quan, L.N., Comin, R., Walters, G., Sabatini, R., Voznyy, O., Hoogland, S., Zhao, Y., Beauregard, E.M., Kanjanaboos, P.: Perovskite energy funnels for efficient light-emitting diodes. Nat. Nanotech. **11**(10), 872–877 (2016). https://doi.org/10.1038/nnano.2016.110
19. Wang, N., et al.: Perovskite light-emitting diodes based on solution-processed self-organized multiple quantum wells. Nat. Photon. **10**, 699 (2016)
20. Xiao, Z., et al.: Efficient perovskite light-emitting diodes featuring nanometre-sized crystallites. Nat. Photon. **11**, 108–115 (2017)
21. Byun, J., et al.: Efficient visible quasi-2D perovskite light-emitting diodes. Adv. Mater. **28**, 7515–7520 (2016)
22. de Quilettes, D.W., et al.: Photoluminescence lifetimes exceeding 8 μs and quantum yields exceeding 30% in hybrid perovskite thin films by ligand passivation. ACS Energy Lett. **1**, 438–444 (2016)
23. Arora, N., et al.: Photovoltaic and amplified spontaneous emission studies of high-quality formamidinium lead bromide perovskite films. Adv. Funct. Mater. **26**, 2846–2854 (2016)
24. Zhang, X., et al.: Enhancing the brightness of cesium lead halide perovskite nanocrystal based green light-emitting devices through the interface engineering with perfluorinated ionomers. Nano Lett. **16**, 1415–1420 (2016)
25. Ling, Y., et al.: Enhanced optical and electrical properties of polymer-assisted all-inorganic perovskites for light-emitting diodes. Adv. Mater. **28**, 8983–8989 (2016)
26. Kulbak, M., Cahen, D., Hodes, G.: How important is the organic part of lead halide perovskite photovoltaic cells? Efficient CsPbBr 3 cells. J. Phys. Chem. Lett. **6**, 2452–2456 (2015)
27. Zhang, L., et al.: Ultra-bright and highly efficient inorganic based perovskite light-emitting diodes. Nat. Commun. **8**, 15640 (2017)
28. Veldhuis, S.A., et al.: Perovskite materials for light-emitting diodes and lasers. Adv. Mater. **28**, 6804–6834 (2016)
29. Li, J., et al.: 50-Fold EQE improvement up to 6.27% of solution-processed all-inorganic perovskite $CsPbBr_3$ QLEDs via surface ligand density control. Adv. Mater. **29**, 1603885 (2017)
30. You, J., et al.: Improved air stability of perovskite solar cells via solution-processed metal oxide transport layers. Nat. Nanotechnol. **11**, 75–81 (2016)
31. Wang, J., et al.: Interfacial control toward efficient and low-voltage perovskite light-emitting diodes. Adv. Mater. **27**, 2311–2316 (2015)
32. Deschler, F., et al.: High photoluminescence efficiency and optically pumped lasing in solution-processed mixed halide perovskite semiconductors. J. Phys. Chem. Lett. **5**, 1421–1426 (2014)
33. Baldo, M.A., Adachi, C., Forrest, S.R.: Transient analysis of organic electrophosphorescence. II. Transient analysis of triplet-triplet annihilation. Phys. Rev. B **62**, 10967–10977 (2000)
34. Luo, Y., Aziz, H., Xu, G., Popovic, Z.D.: Electron-induced quenching of excitons in luminescent materials. Chem. Mater. **19**, 2288–2291 (2007)
35. Popovic, Z.D., Menzel, E.R.: Electric field-induced fluorescence quenching in organic photoconductors. J. Chem. Phys. **71**, 5090–5096 (1979)

36. Gardner, N.F., et al.: Blue-emitting InGaN-GaN double-heterostructure light-emitting diodes reaching maximum quantum efficiency above 200 A/cm^2. Appl. Phys. Lett. **91**, 243506 (2007)
37. Song, J., et al.: Quantum dot light-emitting diodes based on inorganic perovskite cesium lead halides ($CsPbX_3$). Adv. Mater. **27**, 7162–7167 (2015)
38. Li, G., et al.: Highly efficient perovskite nanocrystal light-emitting diodes enabled by a universal crosslinking method. Adv. Mater. **28**, 3528–3534 (2016)
39. Pan, J., et al.: Highly efficient perovskite-quantum-dot light-emitting diodes by surface engineering. Adv. Mater. **28**, 8718–8725 (2016)
40. Chen, Z., et al.: High-performance color-tunable perovskite light emitting devices through structural modulation from bulk to layered film. Adv. Mater. **29**, 1603157 (2017)
41. Yao, E.P., et al.: High-brightness blue and white LEDs based on inorganic perovskite nanocrystals and their composites. Adv. Mater. **29**, 1606859 (2017)
42. Wang, Q., Ren, J., Peng, X.F., Ji, X.X., Yang, X.H.: Efficient sky-blue perovskite light-emitting devices based on ethylammonium bromide induced layered perovskites. ACS Appl. Mater. Inter. **9**, 29901–29906 (2017)
43. Gangishetty, M.K., Hou, S., Quan, Q., Congreve, D.N.: Reducing architecture limitations for efficient blue perovskite light-emitting diodes. Adv. Mater. **30**, 1706226 (2018)
44. Hou, S., Gangishetty, M.K., Quan, Q., Congreve, D.N.: Efficient blue and white perovskite light-emitting diodes via manganese doping. Joule **2**, 2421–2433 (2018)
45. Xing, J., et al.: Color-stable highly luminescent sky-blue perovskite light-emitting diodes. Nat. Commun. **9**, 3541 (2018)
46. Yang, D., et al.: Large-scale synthesis of ultrathin cesium lead bromide perovskite nanoplates with precisely tunable dimensions and their application in blue light-emitting diodes. Nano Energy **47**, 235–242 (2018)
47. Cho, H., et al.: Overcoming the electroluminescence efficiency limitations of perovskite light-emitting diodes. Science **350**, 1222–1225 (2015)
48. Lin, K., et al.: Perovskite light-emitting diodes with external quantum efficiency exceeding 20 per cent. Nature **562**, 245–248 (2018)
49. Cao, Y., et al.: Perovskite light-emitting diodes based on spontaneously formed submicrometre-scale structures. Nature **562**, 249–253 (2018)
50. Zhao, B., et al.: High-efficiency perovskite-polymer bulk heterostructure light-emitting diodes. Nat. Photonics **12**, 783–789 (2018)
51. Chiba, T., et al.: Anion-exchange red perovskite quantum dots with ammonium iodine salts for highly efficient light-emitting devices. Nat. Photon. **12**, 681–687 (2018)
52. Si, J., et al.: Efficient and high-color-purity light-emitting diodes based on in situ grown films of $CsPbX_3$ (X=Br, I) nanoplates with controlled thicknesses. ACS Nano **11**, 11100–11107 (2017)
53. Luo, D., et al.: Enhanced photovoltage for inverted planar heterojunction perovskite solar cells. Science **360**, 1442–1446 (2018)
54. Chen, Z., et al.: Recombination dynamics study on nanostructured perovskite light-emitting devices. Adv. Mater. **30**, 1801370 (2018)
55. Sun, H., Tian, W., Cao, F., Xiong, J., Li, L.: Ultrahigh-performance self-powered flexible double-twisted fibrous broadband perovskite photodetector. Adv. Mater. **30**, 1706986 (2018)
56. Wang, Y., et al.: All-inorganic colloidal perovskite quantum dots: a new class of lasing materials with favorable characteristics. Adv. Mater. **27**, 7101–7108 (2015)
57. Yang, X., Zhang, X., Deng, J., Chu, Z., Jiang, Q., Meng, J., Wang, P., Zhang, L., Yin, Z., You, J.: Efficient green light-emitting diodes based on quasi-two-dimensional composition and phase engineered perovskite with surface passivation. Nat. Commun. **9**(1), 570 (2018). https://doi.org/10.1038/s41467-018-02978-7
58. Congreve, D.N., et al.: Tunable light-emitting diodes utilizing quantum-confined layered perovskite emitters. ACS Photon. **4**, 476–481 (2017)
59. Kumar, S., Jagielski, J., Yakunin, S., Rice, P., Chiu, Y.-C., Wang, M., Nedelcu, G., Kim, Y., Lin, S., Santos, E.J.: Efficient blue electroluminescence using quantum-confined two-dimensional perovskites. ACS Nano **10**(10), 9720–9729 (2016). https://doi.org/10.1021/acsnano.6b05775

60. Liu, Y., Cui, J., Du, K., Tian, H., He, Z., Zhou, Q., Yang, Z., Deng, Y., Chen, D., Zuo, X., Ren, Y., Wang, L., Zhu, H., Zhao, B., Di, D., Wang, J., Friend, R.H., Jin, Y.: Efficient blue light-emitting diodes based on quantum-confined bromide perovskite nanostructures. Nat. Photon. **13**(11), 760–764 (2019). https://doi.org/10.1038/s41566-019-0505-4
61. Liu, T., et al.: Tailoring vertical phase distribution of quasi-two-dimensional perovskite films via surface modification of hole-transporting layer. Nat. Commun. **10**, 878 (2019)
62. Cheng, L., et al.: Multiple-quantum-well perovskites for high-performance light-emitting diodes. Adv. Mater. **32**, 1904163 (2019)
63. Lin, Y., et al.: Suppressed ion migration in low-dimensional perovskites. ACS Energy Lett. **2**, 1571–1572 (2017)
64. Li, C., et al.: Understanding the improvement in the stability of a self-assembled multiple-quantum well perovskite light-emitting diode. J. Phys. Chem. Lett. **10**, 6857–6864 (2019)
65. Leng, M., et al.: Surface passivation of bismuth-based perovskite variant quantum dots to achieve efficient blue emission. Nano Lett. **18**, 6076–6083 (2018)
66. Yin, W.-J., Yang, J.-H., Kang, J., Yan, Y., Wei, S.-H.: Halide perovskite materials for solar cells: a theoretical review. J. Mater. Chem. A **3**, 8926–8942 (2015)
67. Kim, J., Chung, C.-H., Hong, K.-H.: Understanding of the formation of shallow level defects from the intrinsic defects of lead tri-halide perovskites. Phys. Chem. Chem. Phys. **18**, 27143–27147 (2016)
68. Deng, S., et al.: Long-range exciton transport and slow annihilation in two-dimensional hybrid perovskites. Nat. Commun. **11**, 664 (2020)
69. Sapori, D., Kepenekian, M., Pedesseau, L., Katan, C., Even, J.: Quantum confinement and dielectric profiles of colloidal nanoplatelets of halide inorganic and hybrid organic–inorganic perovskites. Nanoscale **8**, 6369–6378 (2016)
70. Xia, P., et al.: A pre-solution mixing precursor method for improving crystallization quality of perovskite film and electroluminescent performance of perovskite light-emitting diodes. Nanoscale **11**, 20847–20856 (2019)
71. Jeong, B., et al.: All-inorganic CsPbI3 perovskite phase-stabilized by poly (ethylene oxide) for red-light-emitting diodes. Adv. Funct. Mater. **28**, 1706401 (2018)
72. Tian, Y., et al.: Highly efficient spectrally stable red perovskite light-emitting diodes. Adv. Mater. **30**, 1707093 (2018)
73. Wu, C., et al.: Improved performance and stability of all-inorganic perovskite light-emitting diodes by antisolvent vapor treatment. Adv. Funct. Mater. **27**, 1700338 (2017)
74. Chu, S., et al.: Large-area and efficient perovskite light-emitting diodes via low-temperature blade-coating. Nat. Commun. **12**, 147 (2021)
75. Kumar, S., et al.: Ultrapure green light-emitting diodes using two-dimensional formamidinium perovskites: achieving recommendation 2020 color coordinates. Nano Lett. **17**, 5277–5284 (2017)
76. Yuan, S., et al.: Self-assembled high quality CsPbBr 3 quantum dot films toward highly efficient light-emitting diodes. ACS Nano **12**, 9541–9548 (2018)
77. Wang, H., et al.: A multi-functional molecular modifier enabling efficient large-area perovskite light-emitting diodes. Joule **4**, 1977–1987 (2020)
78. Zhao, X., et al.: Large-area near-infrared perovskite light-emitting diodes. Nat. Photon. **14**, 215–218 (2020)
79. Kong, L., et al.: Smoothing the energy transfer pathway in quasi-2D perovskite films using methanesulfonate leads to highly efficient light-emitting devices. Nat. Commun. **12**, 1246 (2021)
80. Quan, L.N., et al.: Edge stabilization in reduced-dimensional perovskites. Nat. Commun. **11**, 170–170 (2020)
81. Jiang, Y., et al.: Reducing the impact of Auger recombination in quasi-2D perovskite light-emitting diodes. Nat. Commun. **12**, 336 (2021)
82. Xu, W., et al.: Rational molecular passivation for high-performance perovskite light-emitting diodes. Nat. Photon. **13**, 418–424 (2019)

83. Zhao, X.F., et al.: Opportunities and challenges in perovskite light-emitting devices. ACS Photon. **5**, 3866–3875 (2018)
84. Li, G., et al.: Efficient light-emitting diodes based on nanocrystalline perovksite in a dielectric polymer matrix. Nano Lett. **15**, 2640–2644 (2015)
85. Jiang, Y., Yuan, J., Ni, Y., Yang, J., Wang, Y., Jiu, T., Yuan, M., Chen, J.: Reduced-dimensional α-$CsPbX_3$ perovskites for efficient and stable photovoltaics. Joule **2**(7), 1356–1368 (2018). https://doi.org/10.1016/j.joule.2018.05.004
86. Kim, Y.H., et al.: Multicolored organic/inorganic hybrid perovskite light-emitting diodes. Adv. Mater. **27**, 1248–1254 (2015)
87. Zhang, F.J., et al.: Efficient blue perovskite light-emitting diodes boosted by 2D/3D energy cascade channels. Adv. Funct. Mater. **30**, 2001732 (2020)
88. Hong, X., Ishihara, T., Nurmikko, A.V.: Photoconductivity and electroluminescence in lead iodide based natural quantum well structures. Solid State Commun. **84**, 657–661 (1992)
89. Era, M., Morimoto, S., Tsutsui, T., Saito, S.: Organic–inorganic heterostructure electroluminescent device using a layered perovskite semiconductor $(C_6H_5C_2H_4NH_3)_2PbI_4$. Appl. Phys. Lett. **65**, 676–678 (1994)
90. Yu, J.C., et al.: Improving the stability and performance of perovskite light-emitting diodes by thermal annealing treatment. Adv. Mater. **28**, 6906–6913 (2016)
91. Kim, Y.H., Kim, J.S., Lee, T.W.: Strategies to improve luminescence efficiency of metal-halide perovskites and light-emitting diodes. Adv. Mater. **31**, 1804595 (2019)
92. Lee, H.D., et al.: Efficient Ruddlesden-Popper perovskite light-emitting diodes with randomly oriented nanocrystals. Adv. Funct. Mater. **29**, 1901225 (2019)
93. Kim, D.B., et al.: Uniform and large-area cesium-based quasi-2D perovskite light-emitting diodes using hot-casting method. Adv. Mater. Inter. **7**, 1902158 (2020)
94. Mao, L., Stoumpos, C.C., Kanatzidis, M.G.: Two-dimensional hybrid halide perovskites: principles and promises. J. Am. Chem. Soc. **141**, 1171–1190 (2019)
95. Wang, Z., Wang, F., Zhao, B., Qu, S., Hayat, T., et al.: Efficient two-dimensional tin halide perovskite light-emitting diodes via a spacer cation substitution strategy. J. Phys. Chem. Lett. **11**(3), 1120–1127 (2020). https://doi.org/10.1021/acs.jpclett.9b03565
96. Yuan, S., et al.: Optimization of low-dimensional components of quasi-2D perovskite films for deep-blue light-emitting diodes. Adv. Mater. **31**, 1904319 (2019)
97. Huang, Y., Li, Y., Lim, E.L., Kong, T., Zhang, Y., et al.: Stable layered 2D perovskite solar cells with an efficiency of over 19% via multifunctional interfacial engineering. J. Am. Chem. Soc. **143**(10), 3911–3917 (2021). https://doi.org/10.1021/jacs.0c13087
98. Liu, P., Han, N., Wang, W., Ran, R., Zhou, W., Shao, Z.: High-quality Ruddlesden-Popper perovskite film formation for high-performance perovskite solar cells. Adv. Mater. **33**(10), e2002582 (2021). https://doi.org/10.1002/adma.202002582
99. Kim, H., Pei, M., Lee, Y., Sutanto, A.A., Paek, S., et al.: Self-crystallized multifunctional 2D perovskite for efficient and stable perovskite solar cells. Adv. Funct. Mater. **30**(19), 1910620 (2020). https://doi.org/10.1002/adfm.201910620
100. Lee, J.W., et al.: 2D perovskite stabilized phase-pure formamidinium perovskite solar cells. Nat. Commun. **9**, 3021 (2018)
101. Wang, J., Li, J., Lan, S., Fang, C., Shen, H., et al.: Controllable growth of centimeter-sized 2D perovskite heterostructures for highly narrow dual-band photodetectors. ACS Nano **13**(5), 5473–5484 (2019). https://doi.org/10.1021/acsnano.9b00259
102. Zhou, F., Abdelwahab, I., Leng, K., Loh, K.P., Ji, W.: 2D perovskites with giant excitonic optical nonlinearities for high-performance sub-bandgap photodetection. Adv. Mater. **31**(48), e1904155 (2019). https://doi.org/10.1002/adma.201904155
103. Qin, C., Sandanayaka, A.S.D., Zhao, C., Matsushima, T., Zhang, D., Fujihara, T., Adachi, C.: Stable room-temperature continuous-wave lasing in quasi-2D perovskite films. Nature **585**(7823), 53–57 (2020). https://doi.org/10.1038/s41586-020-2621-1
104. Zhang, H., Liao, Q., Wu, Y., Zhang, Z., Gao, Q., Liu, P., Li, M., Yao, J., Fu, H.: 2D ruddlesden-popper perovskites microring laser array. Adv. Mater. **30**(15), e1706186 (2018). https://doi.org/10.1002/adma.201706186

105. Chou, S., Ma, R., Li, Y., Zhao, F., Tong, K., et al.: Transparent perovskite light-emitting touch-responsive device. ACS Nano **11**(11), 11368–11375 (2017). https://doi.org/10.1021/acsnano.7b05935
106. Liu, Y., Yu, Z., Chen, S., Park, J.H., Jung, E.D., et al.: Boosting the efficiency of quasi-2D perovskites light-emitting diodes by using encapsulation growth method. Nano Energy **80**, 105511 (2021). https://doi.org/10.1016/j.nanoen.2020.105511
107. Meng, F.Y., et al.: Co-interlayer engineering toward efficient green quasi-two-dimensional perovskite light-emitting diodes. Adv. Funct. Mater. **30**, 1910167 (2020)
108. Pang, P.Y., et al.: Rearranging low-dimensional phase distribution of quasi-2D perovskites for efficient sky-blue perovskite light-emitting diodes. ACS Nano **14**, 11420–11430 (2020)
109. Shang, Y., Liao, Y., Wei, Q., Wang, Z., Xiang, B., Ke, Y., Liu, W., Ning, Z.: Highly stable hybrid perovskite light-emitting diodes based on dion-jacobson structure. Sci. Adv. **5**(8), eaaw8072 (2019). https://doi.org/10.1126/sciadv.aaw8072
110. Shi, Y., Wu, W., Dong, H., Li, G., Xi, K., et al.: A strategy for architecture design of crystalline perovskite light-emitting diodes with high performance. Adv. Mater. **30**(25), 1800251 (2018). https://doi.org/10.1002/adma.201800251
111. Xu, L., Li, J., Cai, B., Song, J., Zhang, F., et al.: A bilateral interfacial passivation strategy promoting efficiency and stability of perovskite quantum dot light-emitting diodes. Nat. Commun. **11**(1), 3902 (2020). https://doi.org/10.1038/s41467-020-17633-3
112. Lu, M., Guo, J., Sun, S., Lu, P., Wu, J., et al.: Bright CsPbI3 perovskite quantum dot light-emitting diodes with top-emitting structure and a low efficiency roll-off realized by applying zirconium acetylacetonate surface modification. Nano Lett. **20**(4), 2829–2836 (2020). https://doi.org/10.1021/acs.nanolett.0c00545
113. Jhuang, L., Kumar, G., Chen, F.: Localized surface plasmon resonance of copper nanoparticles improves the performance of quasi-two-dimensional perovskite light-emitting diodes. Dyes Pigm. **188**, 109204 (2021). https://doi.org/10.1016/j.dyepig.2021.109204
114. Zhang, L., Sun, C., He, T., Jiang, Y., Wei, J., et al.: High-performance quasi-2D perovskite light-emitting diodes: from materials to devices. Light Sci. Appl. (2021). https://doi.org/10.1038/s41377-021-00501-0
115. He, Z.F., et al.: High-efficiency red light-emitting diodes based on multiple quantum wells of phenylbutylammonium-cesium lead iodide perovskites. ACS Photon. **6**, 587–594 (2019)
116. Wang, K., Jin, L., Gao, Y., Liang, A., Finkenauer, B.P., et al.: Lead-free organic-perovskite hybrid quantum wells for highly stable light-emitting diodes. ACS Nano **15**(4), 6316–6325 (2021). https://doi.org/10.1021/acsnano.1c00872
117. Liu, M.M., et al.: Suppression of temperature quenching in perovskite nanocrystals for efficient and thermally stable light-emitting diodes. Nat. Photon. **15**, 379–385 (2021)
118. Zhang, Q., et al.: Bifunctional passivation strategy to achieve stable $CsPbBr_3$ nanocrystals with drastically reduced thermal-quenching. J. Phys. Chem. Lett. **11**, 993–999 (2020)
119. Zhang, C.X., et al.: Core/shell perovskite nanocrystals: synthesis of highly efficient and environmentally stable $FAPbBr_3/CsPbBr_3$ for LED applications. Adv. Funct. Mater. **30**, 1910582 (2020)
120. He, Y.R., et al.: Perovskite light-emitting diodes with near unit internal quantum efficiency at low temperatures. Adv. Mater. **33**, 2006302 (2021)
121. Li, X.M., et al.: $CsPbX_3$ quantum dots for lighting and displays: room-temperature synthesis, photoluminescence superiorities, underlying origins and white light-emitting diodes. Adv. Funct. Mater. **26**, 2435–2445 (2016)
122. Pile, D.F.P.: Emitting organically. Nat. Photon. **15**, 635–636 (2021)
123. Woo, S.-J., Kim, J.S., Lee, T.-W.: Characterization of stability and challenges to improve lifetime in perovskite LEDs. Nat. Photon. **15**, 630–634 (2021)
124. Wang, C., Han, D., Wang, J., Yang, Y., Liu, X., et al.: Dimension control of in situ fabricated $CsPbClBr_2$ nanocrystal films toward efficient blue light-emitting diodes. Nat. Commun. **11**(1), 6428 (2020). https://doi.org/10.1038/s41467-020-20163-7
125. Liang, C., Gu, H., Xia, Y., Wang, Z., Liu, X., et al.: Two-dimensional Ruddlesden-Popper layered perovskite solar cells based on phase-pure thin films. Nat. Energy **6**(1), 38–45 (2021). https://doi.org/10.1038/s41560-020-00721-5

Chapter 6
Applications III: Spintronics

Introduction by Editor
Metal halide perovskites exhibit large spin-related effects, which can be separated into two main types: ones common to all or nearly all 2D MHPs, and ones which rely on chiral organic cations to provide an enhanced impact on the spin-dependent electronic structure. Such 2D materials with chiral organic cations are termed chiral MHPs. The 11 papers in this chapter have thus been divided into two groups: the first six papers deal with common effects and the final five papers deal with chiral MHPs. Within each group, the papers are presented chronologically.

The spin effects common to all 2D MHPs include optical selection rules which connect circularly polarized photons to specific spin-dependent valence-to-conduction band transitions, and a large Rashba splitting of the spin states within the electronic band structure. The Rashba effect arises from a large spin-orbit coupling (from e.g. heavy atoms in the material) combined with a broken inversion symmetry due to surfaces or interfaces, and shifts the energy bands of spin up and spin down electrons relative to each other. The Rashba effect also allows for some electrical tunability of the spin properties, as electric fields can determine how strongly the surfaces/interfaces are felt. This strong spin splitting has been predicted theoretically, but also measured experimentally in a number of experiments, including the circular galvanic effect and via a magnetic electroluminescence which changes with an applied electric field. It can affect properties such as hot carrier cooling, as evidenced by a change in the cooling rate obtained through manipulating the spin state of the hot carriers. One application presented here which is made possible through the Rashba effect is a THz radiation source, where the THz radiation is produced by an ultrafast spin current under femtosecond laser excitation.

© The Author(s), under exclusive license to Springer Nature Singapore Pte Ltd. 2024
J. S. Colton and K. R. Hansen (eds.), *Two-Dimensional Metal Halide Perovskites*,
https://doi.org/10.1007/978-981-99-7830-4_6

Spin information can be transferred over surprisingly long times and distances in these materials. Spin lifetimes longer than 1 ns at low temperatures have been measured through time-resolved Faraday rotation. In structures having variable inorganic layer quantum well widths (via the n number), spin information can be transferred from one quantum well to the next as carriers tunnel from a well with larger bandgap to one with a smaller bandgap in the funnel mechanism mentioned in Chap. 5. In one experiment the spin information was able to travel up to 600 nm.

Within the realm of chiral MHPs, the transfer of chirality from the organic molecules to the inorganic lattice can give rise to even more substantial spin splitting in the inorganic-derived conduction band. This is typically evidenced through optical circular dichroism, namely the difference in absorption of one helicity of circularly polarized light vs the other, although circularly-polarized emission can also be seen and used as a chirality transfer marker. In such experiments typically three very similar 2D MHPs are fabricated and results compared: two are made with equivalent organic molecules but having opposite helicities, the third contains a racemic mixture of the two types of molecules. The enhanced effects on electronic spin provide additional possible uses of these materials as spin filters through the chiral-induced spin selectivity effect (CISS), spin LEDs which produce circularly polarized light, and photodetectors for circularly polarized light. These lead to possible enhancements in quantum communications and quantum computing. Although the exact mechanism of chirality transfer is still not fully understood, at least one structural marker in the inorganic layer has been found to correlate with the degree of spin splitting, namely a specific bond angle difference which is connected with asymmetric tilting distortions.

Machine generated summaries

Disclaimer: The auto-summaries have been generated by a recursive clustering algorithm via the Dimensions Auto-summarizer by Digital Science handled by Subject Matter Experts and the external editor. The editor of this book selected which SN content should be auto-summarized and decided its order of appearance. Please be aware that the auto-summaries consist of original sentences, but are not representative of its original paper, since we do not show the full length of the publication. Please note that only published SN content is represented here, and that machine-generated books are still at an experimental stage.

Machine generated keywords: spin, chiral, spintronic, rashba, chirality, symmetry, soc, chiral organic, inversion, spinorbit, magnetic, spintronic device, spinorbit couple, spinpolarize, break

Spin-Polarized Exciton Quantum Beating in Hybrid Organic–Inorganic Perovskites

DOI: https://doi.org/10.1038/nphys4145

Originators: Odenthal, P., Talmadge, W., Gundlach, N. *et al.* Spin-polarized exciton quantum beating in hybrid organic–inorganic perovskites. *Nature Phys* 13, 894–899 (2017). https://doi.org/10.1038/nphys4145

Abstract-Summary

These compounds are also highly promising materials for the field of spintronics due to their large and tunable spin–orbit coupling, spin-dependent optical selection rules, and their predicted electrically tunable Rashba spin splitting.

We demonstrate the optical orientation of excitons and optical detection of spin-polarized exciton quantum beating in polycrystalline films of the hybrid perovskite $CH_3NH_3PbCl_xI_{3-x}$.

These results provide a basic picture of the exciton states in hybrid perovskites, and suggest they hold potential for spintronic applications.

Main

A generic spintronic device consists of spin injection and detection electrodes, a medium connecting the two electrodes in which coherent or polarized spins may be transported and manipulated, and interfaces optimized for efficient spin injection and detection.

The Rashba effect provides an important method for electrical spin manipulation, which is one of the key components in realizing a semiconductor spintronic device.

These advantages, together with their optical accessibility for spin generation and detection, make the perovskites attractive for spintronic applications.

Fundamental understanding of spin-dependent physics in perovskites is still absent, such as accurate description of the spin-polarized states and spin relaxation processes.

Our results suggest that the perovskites are a unique material system in which the usually contradictory long spin lifetime and large SOC coexist; therefore, they hold great promise for spintronic devices.

Sample preparation and characterization

X-ray diffraction data indicate preferred crystal orientation along the (110) direction for the polycrystalline films at room temperature.

Recent absorption measurement at high magnetic field has determined the exciton binding energy to be 16 meV for the low-temperature phase $CH_3NH_3PbI_3$ (refs [1, 2]).

In the temperature range of our experiments, excitons rather than free carriers are the dominant photoexcited species.

The principle of spin dynamics measurements

The unique band structure of $CH_3NH_3PbI_3$ allows optical orientation and detection of spin-polarized excitons.

While $J = 1/2$ and $S = 1/2$ are derived theoretically as the band states for the cubic phase, the structure change to orthorhombic phase can be treated as a perturbation that leaves the symmetry of the electronic states largely intact but folds the band edge states from the R-point to the Γ-point in the Brillouin zone [3].

The perovskites have a reverse ordering of band edge states and similar optical transition rules to GaAs.

Through the same optical transition rules, any polarization between $M = \pm 1$ exciton states can also be detected via optical Faraday rotation that is imparted to a linearly polarized 'probe' beam.

Exciton quantum beating measured by TRFR

For a sufficiently long spin lifetime, the pump-initiated coherent superposition state produces quantum beating in a transverse magnetic field and results in oscillations in TRFR, which is often described as spin precession about the field.

As the Zeeman energies overcome the exchange couplings in increasing magnetic field, the four frequencies reduce down to two frequencies, which in fact correspond to the electron and hole g-factors.

In bulk GaAs, the carrier recombination time is much shorter than the electron spin lifetime and the hole spin lifetime is even shorter; therefore, only resident electron polarization can be observed in TRFR after recombination [4].

The perovskites represent an unusual system in which the exceptional long exciton recombination time and the comparable spin lifetimes of electrons and holes allow the unambiguous observation of spin dynamics of the exciton states.

Energy dependence of the Faraday rotation

We also perform a Hanle effect measurement based on Faraday rotation, which is a steady-state 'pump–probe' measurement.

We repeated the measurements and two Lorentzian fitting for various probe photon energies in the vicinity of the perovskite absorption edge.

The Faraday rotation measures the difference in indices of refraction for LCP and RCP light, which is essentially the energy derivative of the index of refraction spectrum (or second derivative of absorption), if we ignore the energy shift of the two curves.

We note that the measured Faraday rotation spectra reproduce the major zero crossing feature in the second derivative of the absorption spectrum albeit with some broadening.

Temperature dependence

We study the temperature (T) dependence of the spin dynamics in the perovskites, to obtain some insights into the spin relaxation mechanisms.

The spin lifetime τ_s decreases monotonically with temperature, and displays apparently two regions in the log–log plot, with a power-law fitting to $\sim T^{-1.51}$ for the region of T > 15 K. The observed T-dependence here is quite different from a recent experiment [5], where the spin lifetimes in $CH_3NH_3PbI_3$ films were measured above 77 K, showing much shorter spin lifetimes and a slower decrease with temperature.

If we assume momentum scattering is mainly by phonons, a single spin relaxation mechanism cannot satisfactorily describe the experimental data.

The spin relaxation in $CH_3NH_3PbI_3$ may be due to the combination of various mechanisms described above and/or to some novel mechanism yet to be identified [6, 7].

Methods

For the TRFR studies we adopted the degenerate pump–probe scheme, where the laser photon energy was tuned to the band edge (1.64 eV at T = 4 K) for both the pump and probe beams.

A Ti:sapphire laser (SpectraPhysics) was used to generate ~150 fs, 80 MHz pulses, with a spectral width of ~8.5 meV. The pump beam was modulated between RCP and LCP with a photoelastic modulator at 42 kHz, with a typical average intensity of 5.0 W cm^{-2} (fluence 65 nJ cm^{-2}), and the linear polarized probe beam was chopped at 1,237 Hz with a typical average intensity of 0.59 W cm^{-2} (fluence 7.7 nJ cm^{-2}).

The probe beam power was typically kept below 100 μW with a spot size ~50 μm, and the intensity of the pump beam was 10 or more times higher and with a much larger spot size covering the probe spot.

Ultrafast Long-Range Spin-Funneling in Solution-Processed Ruddlesden–Popper Halide Perovskites

DOI: https://doi.org/10.1038/s41467-019-11251-4

This article is licensed under a Creative Commons Attribution 4.0 International License, which permits use, sharing, adaptation, distribution and reproduction in any medium or format, as long as you give appropriate credit to the original author(s) and the source, provide a link to the Creative Commons license, and indicate if changes were made. The images or other third party material in this article are included in the article's Creative Commons license, unless indicated otherwise in a credit line to the material. If material is not included in the article's Creative Commons license and your intended use is not permitted by statutory regulation or exceeds the permitted use, you will need to obtain permission directly from the copyright holder. To view a copy of this license, visit http://creativecommons.org/licenses/by/4.0/.

Originators: Giovanni, D., Lim, J.W.M., Yuan, Z. *et al.* Ultrafast long-range spin-funneling in solution-processed Ruddlesden–Popper halide perovskites. *Nat Commun* 10, 3456 (2019). https://doi.org/10.1038/s41467-019-11251-4

Abstract-Summary

We reveal that solution-processed mixed-phase Ruddlesden–Popper perovskite thin-films transcend the challenges of phonon momentum-scattering that limits spin-transfer in conventional semiconductors.

This highly disordered system exhibits a remarkable efficient ultrafast funneling of photoexcited spin-polarized excitons from two-dimensional (2D) to three-dimensional (3D) phases at room temperature.

This process bypasses the omnipresent phonon momentum-scattering in typical semiconductors with stringent band dispersion, which causes the loss of spin information during thermalization.

Our findings reveal an intriguing family of solution-processed perovskites with extraordinary spin-preserving energy transport properties that could reinvigorate the concepts of spin-information transfer.

Introduction

The future of spintronics hinges on semiconductors that enable efficient room temperature spin generation and transport [8–11].

While the weak spin-orbit coupling (SOC) in organics would theoretically enable long-range transport of spin current [12, 13], it also regrettably imposes huge challenges for spin control [13, 14].

Spintronic studies on low SOC inorganic system, such as graphene [15] and Si [16], have yielded exciting results, where room temperature spin-transport length of ~ 30 μm and ~ 20 μm, respectively, have been demonstrated.

This constraint would limit effective spin transport and could potentially derail halide perovskite's spintronics aspirations.

Moisture resilient mixed-dimensional lead halide perovskites, also known as Ruddlesden–Popper perovskites (RPP) (i.e., $A_2B_{n-1}Pb_nX_{3n+1}$; where A and B are long-chain and small organic cations, respectively; and X = Cl, Br, or I is the halide anion) made their debut in photovoltaics [17] and light emitting diodes (LED) [18].

This novel approach of spin funneling could reinvigorate the concepts of transporting spin information using halide perovskites.

Results

The spin-flip probability due to momentum scattering during thermalization to be α, the observable spin polarization is given by:Here, P_0 is the photoexcited spin polarization and $\Delta E = E_{pump} - E_{probe}$ is the energy difference between the pumping and the probing state (band edge at 1.63 eV), i.e., energy loss due to thermalization.

A weaker counter-circular pump-probe polarization PB signal than its co-circular signal indicates an imbalance of exciton populations in different spin states (i.e., nonzero spin polarization), generated by our circularly polarized-pump beam.

Stronger co- vs. counter-circular PB signal from 1.63 to 2.07 eV spectral region indicates nonzero initial spin polarization of excitons populations in all $n \geq 3$ phases.

That the interphase TA signal in this sample at $\hbar\omega = 1.98$ eV (between n = 3 and n = 4) also shows similar signatures of opposite signal signs for different polarization, as in the standard samples, implying that the same IS-assisted spin-funneling mechanism is also present.

Conclusion

We demonstrated an efficient ultrafast funneling of photoexcited spin-polarized excitons at room temperature in RPP.

Spin-polarized excitons funnel from the low n (2D-like) to high n (3D-like) phases mediated by IS-assisted energy transfer process, which bypasses the phonon momentum scattering during thermalization—thereby preserving the spin information.

Comparable transfer of spin information as for inorganic systems is achieved with perovskites, via this novel energy funneling mechanism.

The discovery of an efficient room-temperature spin funneling platform opens up exciting prospects for perovskite spintronics.

These studies, including this novel concept of spin funneling, could find applications as a new type of perovskite-based spin transistor, which is scattering and defect tolerant toward its spin transport properties.

Our findings highlight the significance of the spin-funneling process in RPP as a novel approach to transcend the strong SOC in perovskites to achieve efficient transfer/transport of spin-polarized excitons/carriers.

Methods

The MAPI precursor solution was prepared by dissolving 1:1 molar stoichiometric ratios of CH_3NH_3I and PbI_2 powders into N,N-dimethylformamide (DMF) with 12.5 wt% concentration.

All films were prepared by spin coating the 20 μL of the precursor solutions on plasma-cleaned quartz substrates, followed by thermal annealing at 10 °C for 30 min.

The 2D/3D graded sample was made by a precursor solution that was prepared by mixing PbI_2 (0.95 mol), $PbCl_2$ (0.05 mol), CH_3NH_3I (0.85 mol), and $C_6H_5C_2H_4NH_3I$ (0.5 mol) in the mixture of N,N-DMF/dimethyl sulfoxide (volume ratio 15:1).

The precursor solution was spin coated on cleaned substrates (4000 rpm for 60 s) directly prior to 100 °C solvent annealing process, continued by films annealing for 10 min.

Linear polarizers together with a Soleil–Babinet compensator and an achromatic quarter-wave plate were used to generate circular polarization for the pump and probe beams, respectively.

Circular Photogalvanic Spectroscopy of Rashba Splitting in 2D Hybrid Organic–Inorganic Perovskite Multiple Quantum Wells

DOI: https://doi.org/10.1038/s41467-019-14073-6

This article is licensed under a Creative Commons Attribution 4.0 International License, which permits use, sharing, adaptation, distribution and reproduction in any medium or format, as long as you give appropriate credit to the original author(s) and the source, provide a link to the Creative Commons license, and indicate if changes were made. The images or other third party material in this article are included in the article's Creative Commons license, unless indicated otherwise in a credit line to the material. If material is not included in the article's Creative Commons license and your intended use is not permitted by statutory regulation or exceeds the permitted use, you will need to obtain permission directly from the copyright holder. To view a copy of this license, visit http://creativecommons.org/licenses/by/4.0/.

Originators: Liu, X., Chanana, A., Huynh, U. *et al.* Circular photogalvanic spectroscopy of Rashba splitting in 2D hybrid organic–inorganic perovskite multiple quantum wells. *Nat Commun* 11, 323 (2020). https://doi.org/10.1038/s41467-019-14073-6

Abstract-Summary

The heavy atoms in 2D-PEPI contribute a large spin-orbit coupling that influences the electronic band structure.

Upon breaking the inversion symmetry, a spin splitting ('Rashba splitting') occurs in the electronic bands.

We have studied the spin splitting in 2D-PEPI single crystals using the circular photogalvanic effect (CPGE).

Whereas the helicity dependent photocurrent with below-gap excitation is due to spin-galvanic effect of the ionized spin-polarized excitons, where spin polarization occurs in the spin-split bands due to asymmetric spin-flip.

Introduction

The three-dimensional (3D) hybrid organic–inorganic halide perovskites (HOIP) with the general formula of AMX_3, where A is an organic cation such as methylammonium (MA), M a divalent metal cation such as Pb^{2+}, and X^- the halide anion such as I^-, have been extensively studied [2–20].

The 2D version of the HOIP are the multilayered Ruddlesden-Popper compounds with the general formula of $A_2A'_{n-1}M_nX_{3n+1}$, with the corner-sharing $[MX_6]^{4-}$ octahedra forming the inorganic perovskite layers separated by the bilayer of interdigitated long chain organic cation A [21, 22], where n indicates the number of inorganic perovskite layers.

Giant Rashba splitting was inferred in a recent spectroscopic study of 2D-PEPI [23] and has since initiated strong interest in this material [24–26].

Unlike the well-studied doped III–V semiconductor MQW structures, where free carriers carry the spin photocurrent [27], it is largely unknown what role, if any, the excitons in 2D-PEPI play in the photogeneration of spin current.

Results

To the CPGE current associated with circularly polarized light, there is also spin-independent photocurrent when the excitation light is linearly polarized (LP, for $\alpha = n(\pi/2)$) [27] This is known as linear photogalvanic effect (LPGE).

The 2D-PEPI crystal is excited with ~50 femtosecond (fs) second-harmonic pulses at 3.06 eV using an amplified Ti-sapphire laser operating at a center wavelength of

810 nm, and the THz emission is collected in transmission geometry (see SM9 in SI).

One possible mechanism to explain these puzzling results is that the helicity-dependent photocurrent at the exciton band is in fact due to the spin-galvanic effect (SGE) rather than the CPGE.

Since the continuum bands in 2D-PEPI are spin splitted due to the Rashba interaction, therefore the exciton-related spin-polarized carriers preferentially occupy one spin sub-band over the other, depending on the light beam helicity.

Discussion

Using circularly polarized light excitation we obtained steady state spin-dependent photocurrent and ultrafast terahertz (THz) emission, which verify the existence of CPGE in 2D-PEPI single crystal multiple quantum wells (MQW).

The CPGE at the interband excitation is a 'smoking gun' proof for Rashba splitting in the continuum bands of 2D-PEPI, which is caused by the large SOC and structural inversion symmetry breaking.

The spin-dependent photocurrent upon exciton excitation is caused by spin-galvanic effect, which also proves the occurrence of Rashba splitting in 2D-PEPI.

Methods

We have grown the 2D hybrid perovskite (2D-PEPI) single crystals on cleaned quartz substrates using the Anti-solvent Vapor-assisted Crystallization (AVC) method as in ref [19] The 2D-PEPI crystals were used for the following measurements: photoluminescence (PL) spectrum, terahertz emission spectroscopy, XRD and SEM microscopy.

CW diode lasers that operate at wavelengths of 405, 447, 486, 520, and 532 nm, respectively, were used to excite the 2D-PEPI single crystal between the two gold electrodes of the device.

Terahertz emission from 2D-PEPI crystals was measured by an electro-optic sampling technique using standard time-domain spectroscopy configuration.

For the photoluminescence (PL) measurement, a 2D-PEPI single crystal was excited using a 30 mW CW laser at 486 nm.

The reflectivity spectrum from 2D-PEPI crystal was measured using Woollam variable angle spectroscopic ellipsometer (VASE) on large area crystals grown on quartz substrates with average thickness 8–13 μm.

Origin of Rashba Spin-Orbit Coupling in 2D and 3D Lead Iodide Perovskites

DOI: https://doi.org/10.1038/s41598-020-61768-8

This article is licensed under a Creative Commons Attribution 4.0 International License, which permits use, sharing, adaptation, distribution and reproduction in any medium or format, as long as you give appropriate credit to the original author(s) and the source, provide a link to the Creative Commons license, and indicate if changes were made. The images or other third party material in this article are included in the article's Creative Commons license, unless indicated otherwise in a credit line to the material. If material is not included in the article's Creative Commons license and your intended use is not permitted by statutory regulation or exceeds the permitted use, you will need to obtain permission directly from the copyright holder. To view a copy of this license, visit http://creativecommons.org/licenses/by/4.0/.

Originators: Pham, M.T., Amerling, E., Luong, H.M. *et al.* Origin of Rashba Spin-Orbit Coupling in 2D and 3D Lead Iodide Perovskites. *Sci Rep* 10, 4964 (2020). https://doi.org/10.1038/s41598-020-61768-8

Abstract-Summary

We studied spin dynamics of charge carriers in the superlattice-like Ruddlesden-Popper hybrid lead iodide perovskite semiconductors, 2D $(BA)_2(MA)Pb_2I_7$ (with MA = CH_3NH_3, and BA = $CH_3(CH_2)_3NH_3$), and 3D $MAPbI_3$ using the magnetic field effect (MFE) on conductivity and electroluminescence in their light emitting diodes (LEDs) at cryogenic temperatures.

We found that the magneto-conductance (MC) magnitude increases monotonically with the emission intensity and saturates at ≈0.05% and 0.11% for the $MAPbI_3$ and $(BA)_2(MA)Pb_2I_7$, respectively.

The magneto-electroluminescence (MEL) response with similar line shapes as the MC response has a significantly larger magnitude, and essentially stays constant at ≈0.22% and ≈0.20% for $MAPbI_3$ and $(BA)_2(MA)Pb_2I_7$, respectively.

The width of the MEL response in those materials linearly increases with increasing the applied electric field, where the Rashba coefficient in $(BA)_2(MA)Pb_2I_7$ is estimated to be about 7 times larger than that in $MAPbI_3$.

Introduction

It has been hypothesized that the strong SOC in the absence of inversion symmetry in OIHPs gives rise to Rashba-type effects, where spin-dependent properties can be manipulated by electric fields [28–31].

A direct comparison of the Rashba effect under the influence of an applied electric field, and the spin interaction in general of 2D and 3D OIHPs has not been experimentally demonstrated.

To the second harmonic generation study of Rashba SOC, the applied magnetic field in the MFE method directly manipulates the spin precession of charge carriers and alters the spin mixing rate between loosely bound singlet and triplet excitons due to their high dielectric screening [32, 33].

In order to quantify these effects, the MFE response of OIHP based LEDs was used to estimate the magnitude of the SOC strength, effective local magnetic field, and Rashba SOC/field as a function of applied electric fields.

Results and discussions

Since the MFE is not observed when the EL is absent and the MEL magnitude is much larger than the MC magnitude, the magnetic field must primarily act on the EL, and MC is just a secondary effect, e.g. from electron-hole pair dissociation.

This implies that the applied electric field might influence the spin mixing processes among the spin sublevels or the intersystem crossing rate between singlet and triplet electron-hole pairs via the Rashba effect [34–36].

Since the spin relaxation time in OIHPs is very fast due to strong SOC, much slower spin precession caused by the hyperfine field cannot cause the spin mixing between the singlet and triplet electron-hole pairs [37].

The basic idea of the exitonic model is that the multiplicity of the singlet and triplet electron-hole pairs/excitons changes with time due to spin dynamics induced by either the Δg mechanism or SOC.

Methods

Prior to any characterization, this precipitate was washed with cold ether and dried in a vacuum oven at 60 °C for 2 h. For the $(BA)_2MAPb_2I_7$ crystals (n = 2), 4.8 mmol (1.1 g) of PbO was dissolved in 10 mL of stabilized HI, heated mildly (60 °C), and stirred until completely dissolved and brilliantly yellow.

The conducting polymer poly(3,4-ethylenedioxythiophene)-poly (styrenesulfonate) (PEDOT) from Ossila was spin coated at 4000 rpm for 60 s on top of the ITO-coated glass substrate yielding a thin PEDOT layer of ≈20 nm thick to provide an efficient hole injecting electrode.

The perovskite solution was spin coated onto the substrate at 5000 rpm for 30 s resulting in films with thicknesses of ≈50 nm and ≈130 nm for $MAPbI_3$ and $(BA)_2(MA)Pb_2I_7$, respectively.

A phenyl-C61-butyric acid methyl ester (PCBM) solution with a concentration of 20 mg/ml in toluene was spin coated directly on the perovskite film to form a thin electron transport layer of ≈20 nm-thick.

Manipulation of Hot Carrier Cooling Dynamics in Two-Dimensional Dion–Jacobson Hybrid Perovskites via Rashba Band Splitting

DOI: https://doi.org/10.1038/s41467-021-24258-7

This article is licensed under a Creative Commons Attribution 4.0 International License, which permits use, sharing, adaptation, distribution and reproduction in any medium or format, as long as you give appropriate credit to the original author(s) and the source, provide a link to the Creative Commons license, and indicate if changes were made. The images or other third party material in this article are included in the article's Creative Commons license, unless indicated otherwise in a credit line to the material. If material is not included in the article's Creative Commons license and your intended use is not permitted by statutory regulation or exceeds the permitted use, you will need to obtain permission directly from the copyright holder. To view a copy of this license, visit http://creativecommons.org/licenses/by/4.0/.

Originators: Yin, J., Naphade, R., Maity, P. *et al.* Manipulation of hot carrier cooling dynamics in two-dimensional Dion–Jacobson hybrid perovskites via Rashba band splitting. *Nat Commun* 12, 3995 (2021). https://doi.org/10.1038/s41467-021-24258-7

Abstract-Summary

Hot-carrier cooling processes of perovskite materials are typically described by a single parabolic band model that includes the effects of carrier-phonon scattering, hot phonon bottleneck, and Auger heating.

We investigated the hot-carrier cooling processes for two slightly different compositions of two-dimensional Dion–Jacobson hybrid perovskites, namely, $(3AMP)PbI_4$ and $(4AMP)PbI_4$ (3AMP = 3-(aminomethyl)piperidinium; 4AMP = 4-(aminomethyl)piperidinium), using a combination of ultrafast transient absorption spectroscopy and first-principles calculations.

In $(4AMP)PbI_4$, upon Rashba band splitting, the spin-dependent scattering of hot electrons is responsible for accelerating hot-carrier cooling at longer delays.

Introduction

Similar to 3D hybrid perovskites, the presence of heavy atoms (e.g., lead) can induce Rashba band splitting because of inversion symmetry breaking [23, 25], making such materials promising candidates for manipulating the spin states in spintronic devices.

There is also a growing interest in exploring applications involving hot-carrier-based optoelectronic and spintronic devices, especially hot-carrier solar cells (HCSCs), using 2D hybrid perovskites due to the deceleration of their hot-carrier cooling rate, which is mainly governed by quantum and dielectric confinement [38–40].

Understanding the mechanisms behind the hot-carrier dynamics in 2D perovskite systems upon Rashba band splitting is critical to realizing their applications in new optoelectronic and spintronic devices.

We explored and deciphered the hot-carrier cooling processes in two slightly different compositions of 2D DJ perovskites, namely, $(3AMP)PbI_4$ and $(4AMP)PbI_4$ (3AMP = 3-(aminomethyl)piperidinium; 4AMP = 4-(aminomethyl)piperidinium), by combining ultrafast time-resolved spectroscopy and first-principles calculations.

Results and Discussion

At a higher pump fluence of 4.0 $\mu J/cm^2$, the initial T_c of the $(3AMP)PbI_4$ film reaches 726 K, and the hot electrons require a longer time to relax to the conduction band edge, giving a time constant of 2.46 ps and an energy loss rate of 0.62 eV/ps.

It can be concluded that the initial hot-carrier cooling in 2D hybrid perovskites is dominated within the subpicosecond time scale (<1 ps) by the emission of optical phonons, while an additional slow cooling process is observed on a timescale of up to hundreds of picoseconds, especially at a high pump fluence, which can be attributed to the hot-phonon bottleneck effect [41, 42].

To confirm and further understand the role of band splitting in the hot-carrier cooling processes, we performed circular polarized transient absorption (CTA) measurements for $(3AMP)PbI_4$ and $(4AMP)PbI_4$ films.

For $(4AMP)PbI_4$, the Rashba band splitting means that the hot-carrier cooling is not only dependent on the interactions between hot electrons and LO-phonons but also affected by spin-dependent scattering of hot electrons [43].

Methods

The detection fiber optics are coupled to a multichannel spectrometer with a CMOS sensor that has a 1.5-nm intrinsic resolution with maximum spectral acquisition rate of 9500 spectra/s. Femtosecond transient absorption measurements were performed for the 2D perovskite films by using a multipass amplified Ti:sapphire laser (800 nm laser pulses with a 7 mJ/pulse; pulse width of ~100 fs with a repetition rate of 1 kHz, Astrella from Coherent) in conjunction with Helios spectrometers.

DFT calculations were carried out for 2D perovskites, $(PMA)_2PbI_4$, $(PEA)_2PbI_4$, $(3AMP)PbI_4$, and $(4AMP)PbI_4$, using the projector-augmented wave (PAW) method as implemented in the VASP code [44, 45].

Uniform Brillouin zone grids of $4 \times 4 \times 2$ k-mesh for $(PMA)_2PbI_4$, $(PEA)_2PbI_4$, and $(4AMP)PbI_4$ and $4 \times 2 \times 2$ for $(3AMP)PbI_4$ were employed, and the plane-wave basis set cutoffs for the wavefunctions and charge density were set at 40 Ry and 300 Ry, respectively.

Coherent Control of Asymmetric Spintronic Terahertz Emission from Two-Dimensional Hybrid Metal Halides

DOI: https://doi.org/10.1038/s41467-021-26011-6

This article is licensed under a Creative Commons Attribution 4.0 International License, which permits use, sharing, adaptation, distribution and reproduction in any medium or format, as long as you give appropriate credit to the original author(s) and the source, provide a link to the Creative Commons license, and indicate if changes were made. The images or other third party material in this article are included in the article's Creative Commons license, unless indicated otherwise in a credit line to the material. If material is not included in the article's Creative Commons license and your intended use is not permitted by statutory regulation or exceeds the permitted use, you will need to obtain permission directly from the copyright holder. To view a copy of this license, visit http://creativecommons.org/licenses/by/4.0/.

Originators: Cong, K., Vetter, E., Yan, L. *et al.* Coherent control of asymmetric spintronic terahertz emission from two-dimensional hybrid metal halides. *Nat Commun* 12, 5744 (2021). https://doi.org/10.1038/s41467-021-26011-6

Abstract-Summary

We report the observation of asymmetric spintronic-THz radiation in Two-Dimensional Hybrid Metal Halides (2D-HMH) interfaced with a ferromagnetic metal, produced by ultrafast spin current under femtosecond laser excitation.

The generated THz radiation exhibits an asymmetric intensity toward forward and backward emission direction whose directionality can be mutually controlled by the direction of applied magnetic field and linear polarization of the laser pulse.

Our work demonstrates the capability for the coherent control of THz emission from 2D-HMHs, enabling their promising applications on the ultrafast timescale as solution-processed material candidates for future THz emitters.

Introduction

Two-Dimensional Hybrid Metal Halides (2D-HMHs), a rising star in synthetic semiconductors prepared by cost-effective, low-temperature solution processing, have recently attracted much research interest [47–49].

The combination of these properties stimulates interest in developing solution-processed, hybrid spin-to-charge transducers on the femtosecond timescale using 2D-HMH materials for a variety of photonic and THz applications [50].

We present the report of asymmetric spintronic-THz emission in 2D-HMHs/ferromagnet heterostructures at room temperature.

Our work shows that 2D-HMHs would be desirable material candidates for spintronic-THz generation and manipulation.

2D-HMH materials may prove superior to current conventional semiconductor materials for THz applications, which require sophisticated deposition approaches that are more susceptible to defects.

Results

Upon linearly polarized fs laser excitation, a superdiffusive spin current is generated in the NiFe layer [51] and, in turn, is injected into the layered 2D-HMH thin film causing the emission of THz radiation via ultrafast spin-to-charge conversion.

While the 1st and 2nd peak groups exhibit different intensities, the +B/−B traces are roughly aligned with each other suggesting an equal intensity of overall THz signals stemming from the same amount of superdiffusive spin current in the 2D-HMH film at opposite magnetic fields.

Both magnetic field and pump polarization-dependent THz measurements have been performed in 3D-HMH-based hybrid spintronic-THz devices, i.e., $CH_3NH_3PbBr_3$/NiFe.

The measured THz electric-field intensity and polarity shows no difference at different angles, confirming that the asymmetric THz radiation neither originate from the magnetic anisotropy of the NiFe nor the crystalline structure of the 2D-HMH film and its relative orientation with respect to the laser polarization axis.

Discussion

We describe the origin of the in-plane momentum shift of Rashba bands due to the concerted actions of an electric dipole along the z-axis and the ordinary Hall effect.

This leads to a rigid in-plane momentum shift of the Rashba bands along the x-axis perpendicular to the applied external magnetic field.

Taking advantage of layered structures in the presence of the injected superdiffusive spin-polarized current, the phase and intensity of THz radiation in 2D-HMH materials can be manipulated by the magnetization state of the adjacent ferromagnet and the pump polarization, offering contactless coherent control of the THz radiation utilizing spintronic toolkits.

Our work will launch a promising testbed for designing a wide variety of low-dimensional HMH materials for future solution-based hybrid spintronic-THz emitter applications, bridging the gap between optoelectronics in HMH materials and THz spintronics.

Methods

2D-HMH films were spin-coated on sapphire substrates in a glove box designed for thin-film preparation.

Following the spin coating procedure, the prepared 2D-HMH thin films were immediately transferred into a glove-box-integrated deposition chamber without exposure to air.

$Ni_{81}Fe_{19}$ and SiO_2 layers were then deposited at room temperature using e-beam evaporation with a chamber base pressure of 1×10^{-7} Torr and evaporation rates of 0.5 and 2.0 A/s, respectively.

The NiFe layer was deposited on a 4×4 mm^2 area of the substrate, while the SiO_2 layer was then coated over the entire 5×5 mm^2 substrate as a capping layer to protect the 2D-HMH thin film during the THz measurement.

The control devices, such as Sapphire/NiFe/Pt, Sapphire/Pt/NiFe, Sapphire/2D-HMH/SiO_2, and Sapphire/NiFe/SiO_2 were prepared following the same protocol.

The weak probe beam is focused on the ZnTe crystal, overlapping with the generated THz pulses spatially and temporally.

Spin Control in Reduced-Dimensional Chiral Perovskites

DOI: https://doi.org/10.1038/s41566-018-0220-6

Originators: Long, G., Jiang, C., Sabatini, R. *et al.* Spin control in reduced-dimensional chiral perovskites. *Nature Photon* 12, 528–533 (2018). https://doi.org/10.1038/s41566-018-0220-6

Abstract-Summary

We report that spin polarization in perovskites can be controlled through chemical design as well as by a magnetic field.

We obtain both spin-polarized photon absorption and spin-polarized photoluminescence in reduced-dimensional chiral perovskites through combined strategies of chirality transfer and energy funnelling.

Three-dimensional perovskites achieve a comparable degree of photoluminescence polarization only under an external magnetic field of 5 T. Our findings pave the way for chiral perovskites as powerful spintronic materials.

Main

We focus here on $\langle n \rangle = 2$ reduced-dimensional chiral perovskites (RDCPs, where $\langle n \rangle$ is the average number of inorganic layers separated by the bulky organic ligands) with the goal of maintaining a large fraction of the chiral organic component in the material and thus increasing the likelihood of chirality transfer.

We began by synthesizing pure chiral 2D perovskites ($\langle n \rangle = 1$) and RDCPs ($\langle n \rangle = 2$) by incorporating chiral organic ligands.

No photoluminescence was observed for the pure chiral 2D $(MBA)_2PbBr_4$ ($\langle n \rangle = 1$) sample, in line with the very poor photoluminescence properties representative of many pure 2D perovskites [52].

Compared with 3D perovskite, the same degree of photoluminescence polarization is obtained only under an external magnetic field of 5 T. RDCPs, which now combine the excellent optoelectronic properties and the function of chirality, would be of great interest in the field of spintronics.

Methods

The Fluorolog system was coupled to a Quanta-Phi integrating sphere with optical fibre bundles for absolute PLQY measurements.

Excitation and emission spectra were measured for three cases: the sample directly illuminated by the beam path in the integrating sphere, the sample indirectly illuminated within the integrating sphere, and the empty sphere.

PLE spectra were measured by monitoring the emission intensity as the excitation wavelength was varied.

A portion of the 1,030 nm fundamental was passed through an optical parametric amplifier (Light Conversion, Orpheus) with the second harmonic of the signal pulse selected for 350 nm light.

A chopper was used to block every other pump pulse, and then the probe pulse was measured by a charge-coupled device (CCD) after being dispersed by a grating spectrograph (Ultrafast, Helios).

Samples were spin-coated on a glass substrate, and were translated at 1 mm s^{-1} during the measurement.

Photoluminescence emission was directed by a multimode optical fibre into a spectrometer (Andor Technology).

Organic-to-Inorganic Structural Chirality Transfer in a 2D Hybrid Perovskite and Impact on Rashba-Dresselhaus Spin-Orbit Coupling

DOI: https://doi.org/10.1038/s41467-020-18485-7

This article is licensed under a Creative Commons Attribution 4.0 International License, which permits use, sharing, adaptation, distribution and reproduction in any medium or format, as long as you give appropriate credit to the original author(s) and the source, provide a link to the Creative Commons license, and indicate if changes were made. The images or other third party material in this article are included in the article's Creative Commons license, unless indicated otherwise in a credit line to the material. If material is not included in the article's Creative Commons license and your intended use is not permitted by statutory regulation or exceeds the permitted use, you will need to obtain permission directly from the copyright holder. To view a copy of this license, visit http://creativecommons.org/licenses/by/4.0/.

Originators: Jana, M.K., Song, R., Liu, H. *et al.* Organic-to-inorganic structural chirality transfer in a 2D hybrid perovskite and impact on Rashba-Dresselhaus spin-orbit coupling. *Nat Commun* 11, 4699 (2020). https://doi.org/10.1038/s41467-020-18485-7

Abstract-Summary

We introduce a structural chirality transfer across the organic–inorganic interface in two-dimensional hybrid perovskites using appropriate chiral organic cations.

The preferred molecular configuration of the chiral spacer cations, R-(+)- or S-(−)-1-(1-naphthyl)ethylammonium and their asymmetric hydrogen-bonding interactions with lead bromide-based layers cause symmetry-breaking helical distortions in the inorganic layers, otherwise absent when employing a racemic mixture of organic spacers.

The ability to break symmetry using chirality transfer from one structural unit to another provides a synthetic design paradigm for emergent properties, including Rashba-Dresselhaus spin-polarization for hybrid perovskite spintronics and related applications.

Introduction

While such chirality transfer phenomena have been mainly established in organic systems, the ability to exploit the influence of chiral organic molecules to structurally modify an extended inorganic lattice and thereby engender emergent properties based on crystal asymmetry remains underexplored.

The inorganic-related chiroptical activity in these already reported HOIP systems is evidently an optically induced phenomenon caused by dipolar interactions with chiral cations [53, 54], and thus, the chain of evidence for structural chirality/asymmetry transfer to the inorganic layers remains incomplete if solely based on chiroptical spectroscopies.

We demonstrate a structural chirality transfer across the organic–inorganic interface in a prototypical 2D HOIP, R-(+)- or S-(−)-1-(1-naphthyl)ethylammonium lead bromide, wherein the enantiopure chiral spacers induce symmetry-breaking helical distortions in the inorganic framework via asymmetric hydrogen-bonding interactions, otherwise absent when employing a racemic mixture of organic spacers.

Results

Such substantial symmetry-breaking distortions in the inorganic framework, owing to the chirality transfer from the chiral NEA^+ spacer cations, are crucial for determining the associated electronic structure and lead to an emergent Rashba–Dresselhaus spin-splitting of electronic bands (discussed below) in chiral R- and S-NPB, otherwise absent when employing a racemic mixture of NEA^+ spacers.

We ascribe the organic-to-inorganic structural chirality transfer in NPB and a lack of detectable transfer in MBPI, at least in part, to the relative strengths of associated H-bonding interactions [e.g., $\Delta H_{(H...Br)} > \Delta H_{(H...I)}$] that mainly determine the templating influence of chiral organic spacers.

While we cannot directly compare bromide/iodide analogs for the same chiral organic cation (from a single-crystal structure standpoint), the key point is that detectable organic-to-inorganic structural chirality transfer does not inevitably follow from the use of a chiral cation within the 2D HOIPs, but additionally relies on the detailed H-bonding interactions that couple the organic and inorganic sublattices.

Discussion

2D HOIPs exhibiting quantum and dielectric confinement of carriers are potential RD candidates [23], in analogy with the inorganic 2DEGs, provided the perovskite layers lack inversion symmetry.

The surrounding chiral cations, the presence of which induces a chiroptical response, need not always entail a significant structural chirality within the inorganic layers; chirality can be structurally transferred to the inorganic layers primarily when mediated by suitably strong hydrogen-bonding interactions with chiral cations, as exemplified by the present chiral NPB systems.

The combination of chiral induced spin selectivity of chiral organic molecules [55–61] and an intrinsic RD spin-splitting in the inorganic framework is a hitherto unexplored aspect unique to 2D chiral HOIPs and might have practical implications in future HOIP-based spintronics.

Further, the structural chirality transfer and ensuing lowering of symmetry within the inorganic sublattice opens up a gateway to other emergent properties such as nonlinear light-matter interactions and piezo-/ferroelectricity, not just in 2D HOIPs but also in related diverse low-dimensional metal halide hybrids.

Methods

Single crystals of racemic-NPB were grown in a similar way from a solution of racemic 1-(1-naphthyl)ethylamine (39 μL, 0.24 mmol) and $PbBr_2$ (45 mg, 0.12 mmol) in 0.5 mL of aq.

HI solution of S-1-(1-naphthyl)ethylamine (0.25 mmol) and PbI_2 (0.125 mmol) from 90 °C to room-temperature in 48 h. The as-obtained pale-yellow, needle-like crystals were filtered, washed with diethyl ether, and vacuum-dried.

To fabricate thin films, 0.2 M solutions of R/S/racemic-NPB single crystals in DMF were spin-cast on glass substrates (precleaned by ultrasonication in IPA for 10 min followed by Ar-O_2 plasma treatment for 10 min) at a spin speed of 3000 rpm for 30 s and annealed at 120 °C for 10 min in an N_2 glove box.

We have: We benchmarked the spin texture functionality within our FHI-aims code against results for several reference systems found in the literature, calculated using other codes.

Structural Descriptor for Enhanced Spin-Splitting in 2D Hybrid Perovskites

DOI: https://doi.org/10.1038/s41467-021-25149-7

This article is licensed under a Creative Commons Attribution 4.0 International License, which permits use, sharing, adaptation, distribution and reproduction in any medium or format, as long as you give appropriate credit to the original author(s) and the source, provide a link to the Creative Commons license, and indicate if changes were made. The images or other third party material in this article are included in the article's Creative Commons license, unless indicated otherwise in a credit line to the material. If material is not included in the article's Creative Commons license and your intended use is not permitted by statutory regulation or exceeds the permitted use, you will need to obtain permission directly from the copyright holder. To view a copy of this license, visit http://creativecommons.org/licenses/by/4.0/.

Originators: Jana, M.K., Song, R., Xie, Y. *et al.* Structural descriptor for enhanced spin-splitting in 2D hybrid perovskites. *Nat Commun* 12, 4982 (2021). https://doi.org/10.1038/s41467-021-25149-7

Abstract-Summary

Two-dimensional (2D) hybrid metal halide perovskites have emerged as outstanding optoelectronic materials and are potential hosts of Rashba/Dresselhaus spin-splitting for spin-selective transport and spin-orbitronics.

Through crystallographic and first-principles studies on a broad array of chiral and achiral 2D perovskites, we demonstrate that a specific bond angle disparity connected with asymmetric tilting distortions of the metal halide octahedra breaks local inversion symmetry and strongly correlates with computed spin-splitting.

This distortion metric can serve as a crystallographic descriptor for rapid discovery of potential candidate materials with strong spin-splitting.

Introduction

When such distortions break the inversion symmetry of the inorganic layer, Rashba/Dresselhaus spin-splitting can emerge due to spin–orbit coupling (SOC) [48, 62–65].

Since strong SOC effects mainly arise in the inorganic component via heavy elements, merely relying on the global space groups without examining the inorganic layer distortions and the resulting local symmetry is, therefore, insufficient to provide a microscopic mechanism of spin-splitting in MHPs.

There have been a few reports with computed spin-splitting values ranging from <<10 meV in some 2D lead-iodide-based MHPs [25, 66] to >200 meV in the chiral $[S/R\text{-}1\text{-}1\text{-}NEA]_2PbBr_4$ [67].

From a materials discovery perspective, a key question is if there exists a simple structural parameter that controls the spin-splitting in MHPs and, consequently, their spin-related properties in applications.

Results

It is not sufficient to merely include chiral spacer cations for the inorganic layers to become chiral (i.e., devoid of inversion and mirror symmetries).

A detectable inversion asymmetry depends on the degree and nature of cation-induced distortions within the inorganic layers.

The global space group alone is, therefore, insufficient as a criterion to predict inversion-asymmetry-induced Rashba/Dresselhaus SOC effects that primarily originate from the inorganic layers.

The actual distortions and resulting local symmetry within the inorganic layer are crucial aspects.

The latter band alignment would have to be validated with a level of theory beyond DFT-PBE but does not impact the conclusions for inorganic-derived bands in this work.

The lack of band dispersion and spin-splitting along the Γ-X path (i.e., along the layer-stacking direction) follows from the confinement and localization of inorganic-derived states due to organic cations acting as insulating barriers.

Discussion

In 2D MHPs, however, noncentrosymmetric global space groups often arise from the molecular organic sublattice and do not necessarily imply strong spin-splitting.

As the SOC effects are principally derived from the inorganic sublattice, the local distortions and the ensuing local asymmetry within the inorganic layers must underlie and control the spin-splitting in 2D MHPs.

While instantaneous structural fluctuations due to phonon modes may quantitatively increase or decrease the spin-splitting of the adiabatic electronic band structure at any given time, the average spin-splitting is expected to retain a clearly defined trend, imparted by the thermal average structure.

For chiral systems, the spin polarization is opposite for opposite organic cation chirality [67], and the ensuing chiral degrees of freedom coupled with a large spin-splitting predicted for some of the current chiral MHPs represent a promising avenue for spin manipulation in prospective spintronic devices.

Methods

Singlecrystals of all the chiral MHPs studied here, except for $[S\text{-}4\text{-}NO_2\text{-}MBA]_2PbBr_4{\cdot}H_2O$ and $[S\text{-}4\text{-}NH_3\text{-}MBA]PbI_4$, were grown by slowly cooling over 60 h a hot (95 °C) aq.

For growing single-crystals of $[S\text{-}4\text{-}NO_2\text{-}MBA]_2PbBr_4{\cdot}H_2O$, a solution of 0.122 mmol of $PbBr_2$ and 0.244 mmol of $S\text{-}4\text{-}NO_2\text{-}MBA{\bullet}HCl$ in 0.5 ml aq.

For growing single-crystals of $[S\text{-}4\text{-}NH_3\text{-}MBA]PbI_4$, a hot solution of 0.245 mmol of $PbI_{2,}$ and 0.49 mmol of $S\text{-}4\text{-}NO_2\text{-}MBA{\bullet}HCl$ in 0.2 ml of aq.

FHI-aims "tight" numerical defaults and k-space grid ($2 \times 4 \times 4$) were employed (except for $[4AMP]PbBr_4$, whose k-space grid was set to ($3 \times 2 \times 5$) due to its specific unit cell).

Electronic band structures suitable for analysis of the impact of SOC were obtained at the PBE level of theory, with a well-benchmarked second-variational non-self-consistent SOC approach [68], and employing FHI-aims "tight" numerical defaults and k-space grid ($2 \times 4 \times 4$).

Control of Light, Spin and Charge with Chiral Metal Halide Semiconductors

DOI: https://doi.org/10.1038/s41570-022-00399-1

Originators: Lu, H., Vardeny, Z.V. & Beard, M.C. Control of light, spin and charge with chiral metal halide semiconductors. *Nat Rev Chem* 6, 470–485 (2022). https://doi.org/10.1038/s41570-022-00399-1

Abstract-Summary

The movement of charges through an oriented chiral medium depends on the spin configuration of the charges, and such systems can be used to control spin populations without magnetic components — termed the chiral-induced spin selectivity (CISS) effect.

Semiconductors are non-magnetic extended systems that allow for the control of charge transport, as well as the absorption and emission of light.

Introducing chirality into semiconductors would enable control over charge, spin and light without magnetic components.

Chiral metal halide semiconductors (MHSs) are hybrid organic–inorganic materials that combine the properties of small chiral organic molecules with those of extended inorganic semiconductors.

Reports of CISS in chiral MHSs have resulted in breakthroughs in our understanding of CISS and in the realization of spin-dependent optoelectronic properties.

Diversity and key structure–property relationships, such as chiral transfer from the organic to the inorganic components, are summarized.

Introduction

The self-assembly of these chiral organic molecules with the inorganic subunits provides an effective strategy to form chiral hybrid organic–inorganic structures, which exhibit properties and functionalities that are absent in the individual components.

The organic molecules transfer chirality to the hybrid structure through interactions with the inorganic component, while the inorganic subunit imparts its optoelectronic semiconducting properties.

Among various hybrid organic–inorganic structures, chiral metal halide semiconductors (MHSs) — which include, but are not limited to, those with perovskite structures — have recently emerged as promising and intriguing materials for spin-dependent optoelectronic devices that can control light, spin and charge.

We then describe the chemical and structural diversity of chiral MHSs and discuss the characteristics that govern the chiral communication between the organic and inorganic subunits, and the way that this affects the chiroptical properties and spin-selectivity.

Background

An investigation in 2017 reported that the metal halide sublattice within a MHS that incorporated chiral organic molecules showed circularly polarized absorption (chiroptical activity), which indicates chirality transfer from the organic to inorganic sublattices [69].

Circularly polarized light leads to 100% spin polarization in MHSs, and the observed circularly polarized absorption in chiral MHSs indicates that they could be used to control spin population.

In 2018, 3% circularly polarized photoluminescence (CPL) was observed at zero magnetic field at 2 K in quasi-2D lead bromide perovskite multilayers incorporated with chiral organic cations [70], suggesting that the photon spin states were controllable with chirality.

In 2019, charge transport in the vertical direction of well aligned 2D chiral lead iodide perovskites was reported to be highly spin-dependent [56], and the spin polarization can be manipulated by the handedness of the organic molecules.

Chemical and structural diversity

The difficulty in forming such structures stems from the lack of suitable chiral organic cations that fit into the A-site of the 3D metal halide perovskite structure (ABX_3, where A^+ is a monovalent cation, B^{2+} is a divalent metal and X^- is a halide).

[71, 72]), 4-fluoro- or chloro- or bromomethylbenzylammonium [73, 74], 1-(1-naphthyl)ethylammonium (NEA^+) [75, 76], 1-cyclohexylethylammonium ($CHEA^+$) [77, 78] and 3,7-dimethyloctylammonium [79], have also been demonstrated to

template inorganic metal halides, including lead iodide, lead bromide, lead chloride, copper chloride and copper bromide, to form stable 2D MHSs with the general formula L_2MX_4 (where L^+ is a monovalent chiral organic cation, M^{2+} is a metal cation and X^- is a halide).

The formation of these 1D chain structures is due to both the branched nature of the chiral organic cations and the connectivity mode of the inorganic motifs.

With chiral organic cations that are not as bulky as MBA^+, such as chiral $MPEA^+$ or 3,7-dimethyloctylammonium, chiral hybrid copper chlorides form 2D structures [79].

Chiral communication between subunits

CD is typically used to characterize the chiroptical activity of chiral systems; however, to quantitatively compare the chiroptical activity across different chiral systems, we need to use the absorption dissymmetry factor, g_{CD}, which is the CD (in millidegrees) normalized by its related absorbance (optical density), A: For luminescent systems, we use the CPL dissymmetry factor, g_{lum}, which is defined by:where I_{left} and I_{right} are, respectively, the photoluminescence emission intensity of left-handed and right-handed circularly polarized light.

Chiral 2D lead bromide nanoplatelets with MPEA as the chiral ligands exhibited a g_{CD} value at the 1s exciton transition as high as 6.5×10^{-3} (ref.

We quantitatively compare the chiroptical activity of molecularly chiral MHSs using the reported g_{CD} values.

For the 1D structure ($MBAPbI_3$) that forms when chiral MBA^+ cations template lead iodide, the measured g_{CD} at the 1s exciton transition is of the order of 10^{-2} (ref.

Chiral-induced spin selectivity

The effective magnetic field will act differently on spins that are oriented parallel or antiparallel to it, depending on the handedness of the chiral molecule.

The spin-filtering effect can be quantified by the spin-polarization, P:where I_+ and I_- are the measured currents at a particular voltage when the magnetic field is pointing up or down, respectively.

This value is much larger than those previously reported for chiral self-assembled monolayer systems, because in the 2D perovskites, the carriers might transfer through multiple chiral layers in a spin-polarized tunnelling process.

This study demonstrated that spin-polarization can be achieved with long-range electron transport and that it does not require an oriented monolayer.

A slightly higher spin-polarization was achieved in the more conductive 2D chiral $(MBA)_2SnI_4$ MHS [61], with a P of +94% and −93% in the $(R\text{-}MBA)_2SnI_4$ and $(S\text{-}MBA)_2SnI_4$ films, respectively.

The spin-polarized current is mostly attributable to the effect of spin-filtering from the oriented chiral organic molecules.

Applications in optospintronics

A heterojunction between a 2D chiral MHS ($(MBA)_2PbI_4$) and MoS_2 was used as a circularly polarized light photodetector, achieving a g_{res} value of 0.09 at a responsivity of 0.45 A W^{-1} and a bias of 3 V (ref [80]).

A 2D chiral MHS, $(R\text{-}MPEA)_2MAPb_2I_7$ with $n = 2$, was used as the photoactive layer to fabricate flexible circularly polarized light photodetectors [81], which displayed a high g_{res} value of 0.2 with a responsivity of 1.1 A W^{-1}.

A vertical device based on 1D chiral $NEAPbI_3$ for the detection of circularly polarized light achieved the highest reported polarization discrimination ratio (R_R/R_L) of 25.4, which corresponds to a g_{res} value as high as 1.85 and a current density of up to 0.28 mA cm^{-2} (ref.

The ability of these devices to discriminate circularly polarized light originates from the dissymmetric absorption of left-handed and right-handed circularly polarized light and the spin-dependent charge transport at the interfaces between the chiral MHS and the other transport layers.

Outlook

The spin-dependent absorption and emission, combined with the semiconducting properties, of chiral MHSs has led to the control of light, spin and charge, enabling circularly polarized light photodetectors and CPL from spin-LEDs.

How chirality is transferred from the organic molecules to the inorganic sublattice remains unclear, and the reported dissymmetry factors of chiral MHSs remain too low for practical applications.

To further advance the unique functionalities of chiral MHSs for the next generation of optoelectronic applications, additional efforts are needed to focus on the chemistry and structural design, understanding of the chirality transfer and CISS mechanism, and device engineering.

The chemical and structural diversity of chiral MHSs provides an excellent platform for studying the fundamental mechanism of CISS, which is not fully understood [82].

The spin-relaxation mechanisms of chiral MHSs remain unclear, requiring investigation of design rules that relate the structure, composition and dimensionality of the chiral MHSs to the spin-relaxation process.

Elucidating the Origin of Chiroptical Activity in chiral 2D Perovskites Through Nano-confined Growth

DOI: https://doi.org/10.1038/s41467-022-31017-9

This article is licensed under a Creative Commons Attribution 4.0 International License, which permits use, sharing, adaptation, distribution and reproduction in any medium or format, as long as you give appropriate credit to the original author(s) and the source, provide a link to the Creative Commons license, and indicate if changes were made. The images or other third party material in this article are included in the article's Creative Commons license, unless indicated otherwise in a credit line to the material. If material is not included in the article's Creative Commons license and your intended use is not permitted by statutory regulation or exceeds the permitted use, you will need to obtain permission directly from the copyright holder. To view a copy of this license, visit http://creativecommons.org/licenses/by/4.0/.

Originators: Ma, S., Jung, YK., Ahn, J. *et al.* Elucidating the origin of chiroptical activity in chiral 2D perovskites through nano-confined growth. *Nat Commun* 13, 3259 (2022). https://doi.org/10.1038/s41467-022-31017-9

Abstract-Summary

Chiral perovskites are being extensively studied as a promising candidate for spintronic- and polarization-based optoelectronic devices due to their interesting spin-polarization properties.

The origin of chiroptical activity in chiral perovskites is still unknown, as the chirality transfer mechanism has been rarely explored.

Our findings suggest that electronic interactions between building blocks should be considered when interpreting the chirality transfer phenomena and designing hybrid materials for future spintronic and polarization-based devices.

Introduction

Via the delicate control of the electronic interaction between the chiral organic molecules and achiral inorganic framework, it is highly expected that the origin of the chirality transfer in chiral OIHPs could be clearly elucidated.

We systematically investigated the effects of electronic interaction between chiral organic spacer cations and inorganic framework on the chiroptical activity of chiral 2D OIHPs by modulating the interaction of the π-electron in chiral organic spacer cations.

The chiroptical activity (i.e., CD and CPPL) of chiral 2D OIHPs dramatically varied depending on the degree of micro-strain due to the varying hydrogen-bonding interactions between chiral organic spacer and achiral inorganic framework, which is originated from the redistributed conformation of benzene rings in the methylbenzylamine (MBA^+) cation.

Our experimental observations combined with theoretical simulation clearly demonstrated that controlling the electronic interaction between the chiral organic spacer cations and inorganic framework becomes a key enabler to design future chiral materials for high-performance spin-polarization-based optoelectronic devices.

Results

To confirm the existence of the micro-strain in the lattice of chiral 2D OIHPs induced by the AAO templates, we conducted PL measurement with the chiral 2D OIHPs grown in different substrate conditions (e.g., planar, 66 nm, 100 nm, and 112 nm pore-sized AAO templates).

Such a coincident tendency (i.e., similar zigzag tendency in calculated micro-strain, PL emission shift as well as excited state splitting) can support that unprecedentedly observed chiroptical conversion behavior in AAO-templated chiral 2D OIHPs results from facilitated chirality transfer phenomena from chiral organic cations (MBA^+) to achiral inorganic framework (lead halide) induced by the micro-strain.

The intensity ratio, I_{FE}/I_{STE}, drastically increased from 1.43 for the planar chiral 2D OIHPs to 3.19 for the 100 nm pore-sized AAO-templated chiral 2D OIHPs, implying that the PbX_4 inorganic framework structure is not distorted by the micro-strain imposed during nanoconfined growth.

Discussion

It can be concluded that the amplified chiroptical activity in AAO-templated chiral 2D OIHPs is more affected by the asymmetric hydrogen-bonding interaction between MBA cations and the inorganic framework rather than the structural distortion in the inorganic framework itself.

While the chirality transfer mechanism in chiral 2D OIHPs has been frequently explained based on the inherent chiral crystal structure formation, the dipolar or electronic interaction mechanism should also be considered to gain a better understanding of chiroptical phenomena and boost the chiroptical activities of chiral 2D OIHPs.

The induced conformational stacking variation promoted the asymmetric hydrogen-bonding interaction between the chiral organic cations and inorganic framework, resulting in excellent asymmetry behavior with g_{CD} of 2.0×10^{-3} and g_{CPPL} of 6.4×10^{-2}.

Our experimental and theoretical calculation results clearly demonstrated that the electronic interaction between the molecular organic spacers and extended inorganic building blocks plays a critical role in interpreting the chirality transfer in chiral 2D OIHPs.

Methods

The resulting solutions were spin coated onto a glass substrate at 3000 rpm for 30 s for planar chiral 2D OIHPs films.

To fabricate the AAO-templated chiral 2D OIHPs thin films, the same precursor solution was used.

The precursor solution was deposited on the AAO substrate with different pore sizes followed by evacuation at 125 Torr for 3 min under a vacuum to promote the infiltration of the precursor solution into the AAO pores.

To measure the circularly polarized response of the chiral 2D OIHPs thin films, the Ti:sapphire laser (450 nm, Mai-tai, Spectra-Physics) was used, and linear light was converted to circularly polarized laser light by using a polarizer and compensator purchased from Newport Co. Based on the Kohn–Sham density functional theory [83], we performed total energy and electronic structure calculations and adopted periodic boundary conditions to represent the extended crystal.

Bibliography

1. D'Innocenzo, V., et al.: Excitons versus free charges in organo-lead tri-halide perovskites. Nat. Commun. **5**, 3586 (2014)
2. Miyata, A., et al.: Direct measurement of the exciton binding energy and effective masses for charge carriers in organic-inorganic trihalide perovskites. Nat. Phys. **11**, 582–587 (2015)
3. Even, J., Pedesseau, L., Jancu, J.-M., Katan, C.: Importance of spin–orbit coupling in hybrid organic/inorganic perovskites for photovoltaic applications. J. Phys. Chem. Lett. **4**, 2999–3005 (2013)
4. Kikkawa, J.M., Awschalom, D.D.: Resonant spin amplification in n-type GaAs. Phys. Rev. Lett. **80**, 4313–4316 (1998)
5. Giovanni, D., et al.: Highly spin-polarized carrier dynamics and ultralarge photoinduced magnetization in $CH_3NH_3PbI_3$ perovskite thin films. Nano Lett. **15**, 1553–1558 (2015)
6. Ma, J., Wang, L.W.: Nanoscale charge localization induced by random orientations of organic molecules in hybrid perovskite $CH_3NH_3PbI_3$. Nano Lett. **15**, 248–253 (2015)
7. March, S.A., et al.: Four-wave mixing in perovskite photovoltaic materials reveals long dephasing times and weaker many-body interactions than GaAs (2016). Preprint at http://arXiv.org/abs/1602.05186v2
8. Dyakonov, M.I.: In: Michel, I.D. (ed.) Spin physics in semiconductors in Springer Series in Solid-State Sciences, vol. 157, chap. 1, pp. 1–28. Springer, Heidelberg, Berlin (2008)
9. Wu, M.W., Jiang, J.H., Weng, M.Q.: Spin dynamics in semiconductors. Phys. Rep. **493**, 61–236 (2010)
10. Žutić, I., Fabian, J., Das Sarma, S.: Spintronics: fundamentals and applications. Rev. Mod. Phys. **76**, 323–410 (2004)
11. Awschalom, D.D., Flatté, M.E.: Challenges for semiconductor spintronics. Nat. Phys. **3**, 153–159 (2007)
12. Dediu, V.A., Hueso, L.E., Bergenti, I., Taliani, C.: Spin routes in organic semiconductors. Nat. Mater. **8**, 707–716 (2009)
13. Boehme, C., Lupton, J.M.: Challenges for organic spintronics. Nat. Nanotechnol. **8**, 612–615 (2013)
14. Sun, D., Ehrenfreund, E., Valy Vardeny, Z.: The first decade of organic spintronics research. Chem. Commun. **50**, 1781–1793 (2014)
15. Gurram, M., Omar, S., van Wees, B.J.: Electrical spin injection, transport, and detection in graphene-hexagonal boron nitride van der Waals heterostructures: progress and perspectives. 2D Mater. **5**, 032004 (2018)
16. Sasaki, T., et al.: Spin transport in nondegenerate Si with a spin MOSFET structure at room temperature. Phys. Rev. Appl. **2**, 034005 (2014)
17. Tsai, H., et al.: High-efficiency two-dimensional Ruddlesden-Popper perovskite solar cells. Nature **536**, 312–316 (2016)
18. Yuan, M., Quan, L.N., Comin, R., Walters, G., Sabatini, R., Voznyy, O., Hoogland, S., Zhao, Y., Beauregard, E.M., Kanjanaboos, P.: Perovskite energy funnels for efficient light-emitting diodes. Nat. Nanotechnol. **11**(10), 872–877 (2016). https://doi.org/10.1038/nnano.2016.110
19. Eames, C., et al.: Ionic transport in hybrid lead iodide perovskite solar cells. Nat. Commun. **6**, 7497 (2015)
20. Hutter, E.M., et al.: Direct-indirect character of the bandgap in methylammonium lead iodide perovskite. Nat. Mater. **16**, 115–120 (2017)
21. Stoumpos, C.C., et al.: Ruddlesden-Popper hybrid lead iodide perovskite 2D homologous semiconductors. Chem. Mater. **28**, 2852–2867 (2016)
22. Du, K.Z., et al.: Two-dimensional lead(II) halide-based hybrid perovskites templated by acene alkylamines: crystal structures, optical properties, and piezoelectricity. Inorg. Chem. **56**, 9291–9302 (2017)
23. Zhai, Y., et al.: Giant Rashba splitting in 2D organic–inorganic halide perovskites measured by transient spectroscopies. Sci. Adv. **3**, e1700704 (2017)

24. Fieramosca, A., et al.: Tunable out-of-plane excitons in 2D single-crystal perovskites. ACS Photon. **5**, 4179–4185 (2018)
25. Yin, J., et al.: Layer-dependent rashba band splitting in 2D hybrid perovskites. Chem. Mater. **30**, 8538–8545 (2018)
26. Chen, X., Lu, H., Li, Z., Zhai, Y., Ndione, P.F., Berry, J.J., Zhu, K., Yang, Y., Beard, M.C.: Impact of layer thickness on the charge carrier and spin coherence lifetime in two-dimensional layered perovskite single crystals. ACS Energy Lett. **3**(9), 2273–2279 (2018). https://doi.org/10.1021/acsenergylett.8b01315
27. Ganichev, S.D., Prettl, W.: Spin photocurrents in quantum wells. J. Phys. Condens. Matter **15**, R935 (2003)
28. Rashba, E.I.: Quantum nanostructures in strongly spin-orbit coupled two-dimensional systems. Phys. Rev. B **86**, 125319 (2012). https://doi.org/10.1103/PhysRevB.86.125319
29. Bychkov, Y.A., Rashba, E.I.: Oscillatory effects and the magnetic susceptibility of carriers in inversion layers. J. Phys. C Solid State Phys. **17**, 6039–6045 (1984). https://doi.org/10.1088/0022-3719/17/33/015
30. Manchon, A., Koo, H.C., Nitta, J., Frolov, S.M., Duine, R.A.: New perspectives for Rashba spin–orbit coupling. Nat. Mater. **14**, 871–882 (2015)
31. Mosconi, E., Etienne, T., De Angelis, F.: Rashba band splitting in organohalide lead perovskites: bulk and surface effects. J. Phys. Chem. Lett. **8**, 2247–2252 (2017). https://doi.org/10.1021/acs.jpclett.7b00328
32. Onoda-Yamamuro, N., Matsuo, T., Suga, H.: Dielectric study of $CH_3NH_3PbX_3$ (X = Cl, Br, I). J. Phys. Chem. Solids **53**, 935–939 (1992)
33. Lin, Q., Armin, A., Nagiri, R.C.R., Burn, P.L., Meredith, P.: Electro-optics of perovskite solar cells. Nat. Photon. **9**, 106 (2014). https://doi.org/10.1038/nphoton.2014.284. https://www.nature.com/articles/nphoton.2014.284#supplementary-information
34. Geng, R., Daugherty, T.T., Do, K., Luong, H.M., Nguyen, T.D.: A review on organic spintronic materials and devices: I. Magnetic field effect on organic light emitting diodes. J. Sci. Adv. Mater. Dev. **1**, 128–140 (2016). https://doi.org/10.1016/j.jsamd.2016.05.002
35. Bloom, F.L., Wagemans, W., Kemerink, M., Koopmans, B.: Separating positive and negative magnetoresistance in organic semiconductor devices. Phys. Rev. Lett. **99**, 257201 (2007). https://doi.org/10.1103/PhysRevLett.99.257201
36. Prigodin, V.N., Bergeson, J.D., Lincoln, D.M., Epstein, A.J.: Anomalous room temperature magnetoresistance in organic semiconductors. Synth. Metals **156**, 757–761 (2006). https://doi.org/10.1016/j.synthmet.2006.04.010
37. Devir-Wolfman, A.H., et al.: Short-lived charge-transfer excitons in organic photovoltaic cells studied by high-field magneto-photocurrent. Nat. Commun. **5**, 4529 (2014). https://doi.org/10.1038/ncomms5529
38. Maity, P., et al.: Layer-dependent coherent acoustic phonons in two-dimensional ruddlesden-popper perovskite crystals. J. Phys. Chem. Lett. **10**, 5259–5264 (2019)
39. Yin, J., et al.: Tuning hot carrier cooling dynamics by dielectric confinement in two-dimensional hybrid perovskite crystals. ACS Nano **13**, 12621–12629 (2019)
40. Wang, T., et al.: Protecting hot carriers by tuning hybrid perovskite structures with alkali cations. Sci. Adv. **6**, eabb1336 (2020)
41. Yang, Y., et al.: Observation of a hot-phonon bottleneck in lead-iodide perovskites. Nat. Photon. **10**, 53–59 (2015)
42. Price, M., et al.: Hot carrier cooling and photo-induced refractive index changes in organic-inorganic lead halide perovskites. Nat. Commun. **6**, 8420 (2015)
43. Todd, S.B., et al.: Detection of Rashba spin splitting in 2D organic-inorganic perovskite via precessional carrier spin relaxation. APL Mater. **7**, 081116 (2019)
44. Kresse, G., Hafner, J.: Ab-initio molecular-dynamics for open-shell transition-metals. Phys. Rev. B **48**, 13115–13118 (1993)
45. Kresse, G., Furthmüller, J.: Efficient iterative schemes for ab initio total-energy calculations using a plane-wave basis set. Phys. Rev. B **54**, 11169–11186 (1996)

46. Wu, Q., Zhang, X.-C.: Free-space electro-optic sampling of terahertz beams. Appl. Phys. Lett. **67**, 3523–3525 (1995)
47. Wang, J., et al.: Spin-optoelectronic devices based on hybrid organic-inorganic trihalide perovskites. Nat. Commun. **10**, 129 (2019)
48. Kepenekian, M., et al.: Rashba and Dresselhaus effects in hybrid organic–inorganic perovskites: from basics to devices. ACS Nano **9**, 11557–11567 (2015)
49. Sun, D., et al.: Surface-enhanced spin current to charge current conversion efficiency in $CH_3NH_3PbBr_3$-based devices. J. Chem. Phys. **151**, 174709 (2019)
50. Chanana, A., Liu, X., Zhang, C., Vardeny, Z.V., Nahata, A.: Ultrafast frequency-agile terahertz devices using methylammonium lead halide perovskites. Sci. Adv. **4**, eaar7353 (2018)
51. Eschenlohr, A., et al.: Ultrafast spin transport as key to femtosecond demagnetization. Nat. Mater. **12**, 332–336 (2013)
52. Xing, G., et al.: Transcending the slow bimolecular recombination in lead-halide perovskites for electroluminescence. Nat. Commun. **8**, 14558 (2017)
53. Gao, X., Han, B., Yang, X., Tang, Z.: Perspective of chiral colloidal semiconductor nanocrystals: opportunity and challenge. J. Am. Chem. Soc. **141**, 13700–13707 (2019)
54. Govorov, A.O., Fan, Z., Hernandez, P., Slocik, J.M., Naik, R.R.: Theory of circular dichroism of nanomaterials comprising chiral molecules and nanocrystals: plasmon enhancement, dipole interactions, and dielectric effects. Nano Lett. **10**, 1374–1382 (2010)
55. Naaman, R., Paltiel, Y., Waldeck, D.H.: Chiral molecules and the spin selectivity effect. J. Phys. Chem. Lett. **11**, 3660–3666 (2020)
56. Lu, H., et al.: Spin-dependent charge transport through 2D chiral hybrid lead-iodide perovskites. Sci. Adv. **5**, eaay0571 (2019)
57. Long, G., et al.: Chiral-perovskite optoelectronics. Nat. Rev. Mater. **5**, 423–439 (2020)
58. Abendroth, J.M., et al.: Spin selectivity in photoinduced charge-transfer mediated by chiral molecules. ACS Nano **13**, 4928–4946 (2019)
59. Ghosh, K.B., et al.: Controlling chemical selectivity in electrocatalysis with chiral CuO-coated electrodes. J. Phys. Chem. C **123**, 3024–3031 (2019)
60. Abendroth, J.M., et al.: Spin-dependent ionization of chiral molecular films. J. Am. Chem. Soc. **141**, 3863–3874 (2019)
61. Lu, H., et al.: Highly distorted chiral two-dimensional tin iodide perovskites for spin polarized charge transport. J. Am. Chem. Soc. **142**, 13030–13040 (2020)
62. Liu, X., et al.: Circular photogalvanic spectroscopy of Rashba splitting in 2D hybrid organic–inorganic perovskite multiple quantum wells. Nat. Commun. **11**, 323 (2020)
63. Martin Schlipf, F.G.: Dynamic Rashba-Dresselhaus effect (2020). Preprint at https://arxiv.org/abs/2004.10477
64. Wang, F., et al.: Switchable Rashba anisotropy in layered hybrid organic–inorganic perovskite by hybrid improper ferroelectricity. NPJ Comput. Mater. 6, 183 (2020)
65. Niesner, D., et al.: Structural fluctuations cause spin-split bands in $(CH_3NH_3)PbI_3$: experimental evidence from circular photogalvanic effect. PNAS **115**, 9509–9514 (2018)
66. Schmitt, T., et al.: Control of crystal symmetry breaking with halogen-substituted benzylammonium in layered hybrid metal-halide perovskites. J. Am. Chem. Soc. **142**, 5060–5067 (2020)
67. Jana, M.K., et al.: Organic-to-inorganic structural chirality transfer in a 2D hybrid perovskite and impact on Rashba-Dresselhaus spin-orbit coupling. Nat. Commun. **11**, 4699 (2020)
68. Huhn, W.P., Blum, V.: One-hundred-three compound band-structure benchmark of post-self-consistent spin-orbit coupling treatments in density functional theory. Phys. Rev. Mater. **1**, 033803 (2017)
69. Ahn, J., et al.: A new class of chiral semiconductors: chiral-organic-molecule-incorporating organic–inorganic hybrid perovskites. Mater. Horiz. **4**, 851–856 (2017)
70. Long, G., et al.: Spin control in reduced-dimensional chiral perovskites. Nat. Photon. **12**, 528–533 (2018)
71. Yuan, C., et al.: Chiral lead halide perovskite nanowires for second-order nonlinear optics. Nano Lett. **18**, 5411–5417 (2018)

72. Huang, P.-J., Taniguchi, K., Miyasaka, H.: Bulk photovoltaic effect in a pair of chiral–polar layered perovskite-type lead iodides altered by chirality of organic cations. J. Am. Chem. Soc. **141**, 14520–14523 (2019)
73. Yang, C.-K., et al.: The first 2D homochiral lead iodide perovskite ferroelectrics: [R- and S-1-(4-chlorophenyl)ethylammonium]2PbI_4. Adv. Mater. **31**, 1808088 (2019)
74. Zhou, C., et al.: Photoluminescence spectral broadening, chirality transfer and amplification of chiral perovskite materials (R-X-p-mBZA)2PbBr_4 (X = H, F, Cl, Br) regulated by van der Waals and halogen atoms interactions. Phys. Chem. Chem. Phys. **22**, 17299–17305 (2020)
75. Zhu, L.-L., et al.: Stereochemically active lead chloride enantiomers mediated by homochiral organic cation. Polyhedron **158**, 445–448 (2019)
76. Jana, M.K., et al.: Organic-to-inorganic structural chirality transfer in a 2D hybrid perovskite and impact on Rashba-Dresselhaus spin–orbit coupling. Nat. Commun. **11**, 4699 (2020)
77. Lemmerer, A., Billing, D.G.: Inorganic–organic hybrids incorporating a chiral cyclic ammonium cation. J. South Afr. J. Chem. **66**, 263–272 (2013)
78. Hu, Y., et al.: A chiral switchable photovoltaic ferroelectric 1D perovskite. Sci. Adv. **6**, eaay4213 (2020)
79. Takahashi, M., Hoshino, N., Sambe, K., Takeda, T., Akutagawa, T.: Dynamics of chiral cations in two-dimensional CuX4 and PbX_4 perovskites (X = Cl and Br). Inorg. Chem. **59**, 11606–11615 (2020)
80. Ma, J., et al.: Chiral 2D perovskites with a high degree of circularly polarized photoluminescence. ACS Nano **13**, 3659–3665 (2019)
81. Wang, L., et al.: A chiral reduced-dimension perovskite for an efficient flexible circularly polarized light photodetector. Angew. Chem. Int. Ed. **59**, 6442–6450 (2020)
82. Evers, F., et al.: Theory of chirality induced spin selectivity: progress and challenges. Adv. Mater. **34**, 2106629 (2022)
83. Kohn, W., Sham, L.J.: Self-consistent equations including exchange and correlation effects. Phys. Rev. **140**, A1133 (1965)

Chapter 7
Applications IV: The More Distant Future

Introduction by Editor

The articles in Chaps. 4, 5, and 6 cover the most widely discussed applications of 2D MHPs as solar cells, optical detectors, LEDs, and spin-selective devices. Yet, the landscape of possibilities for 2D MHPs stretches beyond these established applications. This chapter is focused on highlighting the unconventional, more distant-future intriguing applications of 2D MHPs. The articles in this chapter are presented in chronological order and cover many unique potential applications for 2D MHPs, namely radiation detectors, terahertz (THz) modulators, graphene heterostucture transistors, and polaritonic devices.

As put forward by the articles in this chapter, 2D MHPs show promise in the space of gamma-ray detection, hard X-ray detection, beta-decay monitoring, and possibly neutron detection as well. In the realm of X-ray and gamma-ray detection, challenges persist with traditional scintillators due to self-absorption effects. Preliminary research suggests that by leveraging the strain applied to 2D MHPs, it may be possible to mitigate self-absorption without diminishing the radiation response speed. The defect tolerance of MHPs is also highly advantageous in high-energy radiation detection since traditional scintillators are defect sensitive. However, the stability of 2D MHP detectors in conditions of intense radiation and strong electric fields needs to be further investigated to determine the viability of MHP radiation detectors, especially considering that some studies have shown that 3D MHPs experience spectrum degradation upon exposure to a large amount of radiation.

There also is growing interest in using 2D MHPs as modulators of THz radiation. Excitonic absorption below the band gap gives rise to a photo-induced THz absorption effect which can be exploited to control the intensity and the wavelength of the THz signal. Both of these effects are highly sought after for next-generation THz-based communication systems.

© The Author(s), under exclusive license to Springer Nature Singapore Pte Ltd. 2024
J. S. Colton and K. R. Hansen (eds.), *Two-Dimensional Metal Halide Perovskites*,
https://doi.org/10.1007/978-981-99-7830-4_7

Templating 2D MHP layers with graphene shows promise for various optoelectronic applications such as next-generation field-effect transistors. The energy barrier for charge injection across the perovskite-graphene interface has been demonstrated to be even lower than a perovskite-gold interface. Field-effect transistors designed around the 2D MHP-graphene junction show high electron mobilities and these devices may prove useful if optimized further.

Finally, the large exciton binding energy and excellent light-emitting properties of 2D MHPs makes these materials ideal candidates for room-temperature exciton-polariton devices. Such devices include low threshold polariton lasers, polariton logic devices, and neuromorphic computing. While achieving polariton condensation has been challenging due to the broad absorption features of MHPs at room temperature, epitaxial growth of the materials will likely help improve their quality and lead to greater viability in this space.

Machine generated summaries

Disclaimer: The auto-summaries have been generated by a recursive clustering algorithm via the Dimensions Auto-summarizer by Digital Science handled by Subject Matter Experts and the external editor. The editor of this book selected which SN content should be auto-summarized and decided its order of appearance. Please be aware that the auto-summaries consist of original sentences, but are not representative of its original paper, since we do not show the full length of the publication. Please note that only published SN content is represented here, and that machine-generated books are still at an experimental stage.

Machine generated keywords: radiation, ray, detection, nuclear, detector, xray, imaging, security, sensitive, highenergy, roomtemperature, atomic, create, type, graphene

Two-Dimensional Halide Perovskite as β-Ray Scintillator for Nuclear Radiation Monitoring

DOI: https://doi.org/10.1038/s41467-020-17114-7

This article is licensed under a Creative Commons Attribution 4.0 International License, which permits use, sharing, adaptation, distribution and reproduction in any medium or format, as long as you give appropriate credit to the original author(s) and the source, provide a link to the Creative Commons license, and indicate if changes were made. The images or other third party material in this article are included in the article's Creative Commons license, unless indicated otherwise in a credit line to the material. If material is not included in the article's Creative Commons license and your intended use is not permitted by statutory regulation or exceeds the permitted use, you will need to obtain permission directly from the copyright holder. To view a copy of this license, visit http://creativecommons.org/licenses/by/4.0/.

Originators: Yu, D., Wang, P., Cao, F. *et al.* Two-dimensional halide perovskite as β-ray scintillator for nuclear radiation monitoring. *Nat Commun* 11, 3395 (2020). https://doi.org/10.1038/s41467-020-17114-7

Abstract-Summary

β-ray monitoring is a crucial method, but commercial organic scintillators for β-ray detection suffer from high temperature failure and irradiation damage.

We report a type of β-ray scintillator with good thermotolerance and irradiation hardness based on a two-dimensional halide perovskite.

Besides, the two-dimensional halide perovskite β-ray scintillator also overcomes the notorious intrinsic water instability, and benefits from low-cost aqueous synthesis along with superior waterproofness, thus paving the way towards practical application.

Introduction

Nuclear reaction mainly releases four types of high-energy rays: α-ray, β-ray, γ-ray, and neutron ray.

Scintillator-based detectors that convert high-energy rays into visible light are facile, efficient, and cost-effective, hence are widely used for radiation detection.

Organic scintillators (including single-crystal, liquid, and plastic types) are mainly used for β-ray detection.

A type of β-ray scintillator based on two-dimensional (2D) HPs is demonstrated.

Comprehensive compositional engineering was carried out towards β-ray sensing: (1) a series of bulky organic cations in 2D HPs were explored to enhance the capturing of β particles, therefore significantly increasing the scintillation response to β-ray; (2) Extrinsic manganese (Mn(II)) dopants were incorporated, which not only serve as extra emitting centers, but also play as emission shifter to eliminate the self-absorption caused photon loss.

Results

To evaluate the effect of composition on the scintillation response toward β-ray, the so-called "zero-dimensional" (0D) HP $CsPbBr_3@Cs_4PbBr_6$ is used for reference [1, 2].

After β particles are absorbed by 2D HPs, the incorporation of light elements also enhances the scintillation response by suppressing energy allocation on useless signals.

To further evaluate the irradiation stability of the 2D HPs as β-ray scintillators, a test was conducted on STA_2PbBr_4:Mn as an example.

Proof-of-concept radiation test is then demonstrated to show that the 2D HP scintillator offers effective and highly discernable response to β-ray.

To get an idea of how small the detection limit is, we refer to the file of "WS 533-2017 protection requirements on nuclear medicine patient", it states that the irradiation activity must be $<2 \times 10^8$ Bq (pure β-ray by ^{89}Sr in medicine) before the patient can be discharged, it means even in such an unoptimized measuring setup, the low-detection limit 2D HP scintillator already meets the requirement of medical diagnosis.

Discussion

A type of β-ray scintillator based on 2D HPs is demonstrated.

The design principle of a high-performance β-ray scintillator is discussed, according to which comprehensive composition engineering is carried out.

It is believed that this work will open new possibilities of HPs towards high-performance β-ray detection for nuclear radiation monitoring.

Methods

STA was dissolved in 300 ml acetic acid, then HI/water solution was added under vigorous stirring.

The (STA)I was immediately extracted by 10,000 × 3 min centrifugation and washed using pure acetic acid, the centrifugation-washing cycle was repeated several times until the supernatant became transparent and the (STA)I precipitates became totally white.

2:1 (STA)Br and $PbBr_2$ were dissolved in DMSO to form precursor with a high concentration of 0.5 M. In all, 0.5 ml of the precursor was injected into 20 ml deionized water under vigorous stirring.

Taking BA_2PbBr_4 for example, 2:1 (BA)Br and $PbBr_2$ were dissolved in DMSO to form precursor with a high concentration of 0.5 M. In total, 0.5 ml of the precursor was injected into 20 ml toluene under vigorous stirring.

Hybrid Halide Perovskite Neutron Detectors

DOI: https://doi.org/10.1038/s41598-021-95586-3

This article is licensed under a Creative Commons Attribution 4.0 International License, which permits use, sharing, adaptation, distribution and reproduction in any medium or format, as long as you give appropriate credit to the original author(s) and the source, provide a link to the Creative Commons license, and indicate if changes were made. The images or other third party material in this article are included in the article's Creative Commons license, unless indicated otherwise in a credit line to the material. If material is not included in the article's Creative Commons license and your intended use is not permitted by statutory regulation or exceeds the permitted use, you will need to obtain permission directly from the copyright holder. To view a copy of this license, visit http://creativecommons.org/licenses/by/4.0/.

Originators: Andričević, P., Náfrádi, G., Kollár, M. *et al.* Hybrid halide perovskite neutron detectors. *Sci Rep* 11, 17159 (2021). https://doi.org/10.1038/s41598-021-95586-3

Abstract-Summary

Interest in fast and easy detection of high-energy radiation (x-, γ-rays and neutrons) is closely related to numerous practical applications ranging from biomedicine and industry to homeland security issues.

Crystals of hybrid halide perovskite have proven to be excellent detectors of x- and γ-rays, offering exceptionally high sensitivities in parallel to the ease of design and handling.

We demonstrate that by assembling a methylammonium lead tri-bromide perovskite single crystal ($CH_3NH_3PbBr_3$ SC) with a Gadolinium (Gd) foil, one can very efficiently detect a flux of thermal neutrons.

The neutrons absorbed by the Gd foil turn into γ-rays, which photo-generate charge carriers in the $CH_3NH_3PbBr_3$ SC.

The induced photo-carriers contribute to the electric current, which can easily be measured, providing information on the radiation intensity of thermal neutrons.

Introduction

Pertinent questions arise: (1) Could perovskite-based devices detect these leaked neutrons? (2) Could one use them for energy production? (3) What would be the effect of the neutrons on the perovskite radiation detectors in the reactor?

Doping these detectors with elements that could increase the neutron capture efficiency, for example ^{10}B enriched hBN neutron detectors [3], should be the way for increasing the detection efficiency of semiconductor-based devices.

The neutrons are converted to secondary radiation directly by the $MAPbBr_3$ crystal due to nuclear interactions, for example via the prompt gamma production (n,γ), which ultimately generates charge carriers.

While, similarly to its bromine counterpart, the $MAPbI_3$ SC—based device demonstrated accurate and stable detection of the neutron beam, the detector based on the $MAPbCl_3$ SC did not show sufficient sensitivity.

Experimental section

The $MAPbBr_3$ SCs used in the fabrication of the detector devices were grown using the inverse temperature crystallization method [4].

To test the ability of neutron detection by the perovskite, the fabricated detectors were positioned in a direct beam of cold-thermal neutrons from 0.9 Å or ~ 101 meV to 15 Å or ~ 0.36 meV at the Polref reflectometer [5] of the ISIS Neutron and Muon Source Facility.

The Polref reflectometer was positioned at Target Station 2, thus neutron packages arrived at 10 Hz pulse frequency, after being moderated in a coupled grooved cold H_2/CH_4 moderator cell.

The detectors were biased in the 0–10 V range (Keithley 2400 source meter), depending on the crystal size, while measuring the current through the device.

Results and discussion

Calculations give that the 10^7 n/cm^2/s neutron flux is converted into 2×10^6 γ/cm^2/s photon flux in the $MAPbBr_3$ SC device.

To boost the low-energy gamma radiation and the overall detector efficiency a converter layer was positioned in the neutron-beam in front of the perovskite SC.

Neutrons from the beam interact with the layer, generating Gamma rays, which are in turn detected by the perovskite SC.

To test if the self-healing mechanism will still take place in the case of thermal neutrons, i.e. if we would see a decrease in device performance, a SC $MAPbBr_3$ detector was irradiated for 15 h under the direct neutron beam.

The long-term measurement was then repeated after inserting the Gd-foil converter in front of the $MAPbBr_3$ SC, to test the operational stability of our neutron beam detector.

Conclusion and outlook

Neutron detectors were fabricated using $MAPbBr_3$ SCs as the sensing material.

The high-sensitivity device architecture consisted of a converter layer, a gadolinium foil or a borated polyethylene sheet, which converted the incoming thermal neutrons to prompt gamma radiation.

These gamma rays were then detected by the highly sensitive perovskite single crystals equipped with graphite spray electrodes.

The simplicity of the detection (by photocurrent measurement), low-cost fabrication process and stability under radiation makes these methylammonium lead tri-halide perovskites a good candidate for the next generation of neutron detectors.

Methods

The MCNP model contains the geometry which in this case consists of a Gd converter foil ($0.025 \times 5 \times 5$ cm^3), a $MAPbBr_3$ SC ($2 \times 2 \times 1$ cm^3, the actual size of the crystal used in the experiments) with graphite coating around two-third of it (0.05 cm thick) and an aluminium platform ($11.505 \times 20 \times 4.495$ cm^3) beneath the SC.

Calculations were repeated without converter layers too to calculate the neutron to gamma conversion inside the crystal.

The obtained prompt gamma background flux at the sample position was 1.38×10^4 p/cm^2/s. Increasing the guide opening to the size of the opening of the shutter (further overestimation) resulted in a higher gamma flux of 2.08×10^5 p/cm^2/s. These overestimated photon fluxes are far smaller than the converted fluxes even without converter layers.

Detecting Ionizing Radiation Using Halide Perovskite Semiconductors Processed Through Solution and Alternative Methods

DOI: https://doi.org/10.1038/s41566-021-00909-5

Originators: He, Y., Hadar, I. & Kanatzidis, M.G. Detecting ionizing radiation using halide perovskite semiconductors processed through solution and alternative methods. *Nat. Photon.* 16, 14–26 (2022). https://doi.org/10.1038/s41566-021-00909-5

Abstract-Summary

The direct detection of high-energy radiation such as X-rays and γ-rays by semiconductors at room temperature is a challenging proposition that requires remarkably pure and nearly perfect crystals.

Metal halide perovskites, which combine exceptional optoelectronic properties, versatile chemistry and simple synthesis, are challenging traditional approaches for the development of novel semiconductors for detecting hard radiation.

We discuss the relevant physical properties, promising materials, fabrication techniques and device architectures for high-performance, low-cost detectors by targeting next-generation semiconductors for radiation detection.

Main

The direct detection of X- and γ-rays by semiconductor materials at room temperature is of great interest, as such devices are anticipated to offer unparalleled performance in terms of image quality, energy resolution and system volume.

These applications are primarily based on two semiconductor materials: amorphous Se (a-Se) films in large-area radiography imaging and bulk $Cd_{1-x}Zn_xTe$ (CZT, $x \approx 0.1$) single crystals in homeland security and nuclear medical imaging [6, 7].

This Review focuses on the application of MHPs as semiconductor detectors for high-energy X- and γ-ray detection.

We then review the MHP semiconductors used for X- and γ-ray spectroscopy in the single-photon mode.

We anticipate that this Review will direct the attention of other members of the MHP community towards materials development for X- and γ-ray detection and encourage researchers focused on detectors to explore and further develop advanced devices based on MHPs.

Application-oriented criteria for material properties

This mechanism produces several excited charges after an interaction with a single high-energy photon, which can be detected.

When ε_{pair} decreases, more charges can form in a single interaction, which may increase the sensitivity and improve the energy resolution; however, a lower bandgap can also result in a higher dark current, causing additional electrical noise.

The detection of high-energy photons can be divided into two primary motifs: the charge-integration and single-photon modes.

In a thin-film transistor-based X-ray imager that is operated in the charge-integration mode [8], a dark current is inevitably added to the signal, resulting in an increase in electronic noise and a reduction in the signal-to-noise ratio.

MHP detectors for these applications should be based on high-quality semiconductors that could show superior optoelectronic properties, including an exceptional energy resolution when compared with CZT and inorganic scintillators.

Relationship between crystal structure and detector performance in MHPs

In 3D MHPs that are free of spacers, the carrier mobility has been determined to range from approximately 10 to several hundred $cm^2\ V^{-1}\ s^{-1}$.

For the 3D MHPs a plausible inverse correlation is observed between the dark current and X-ray sensitivity; in all cases, the dark current in 3D MHPs is still on the order of 1 nA cm^{-2} or more owing to the limited resistivity on the order of 10^9 Ω cm.

2D MHPs with a higher resistivity of up to 10^{11} Ω cm showed a dramatically reduced dark current, approaching the order of 100 pA cm^{-2}, which may satisfy the current requirement for projection radiography.

Despite the apparent 3D connectivity in the double perovskite $Cs_2AgBiBr_6$, its physical behaviour resembles that of low-dimensional MHPs, featuring a dark current that is comparable to 2D counterparts (close to 100 pA cm^{-2}) and moderate X-ray sensitivity [9–12].

Synthesis of perovskite semiconductor detectors, solution processing and alternative methods

As the standard spin-coating process used for photovoltaic and light-emitting devices usually yields MHP films with thicknesses of only a few hundred nanometres [13], alternative fabrication processes are being pursed to grow the considerably thicker films that are required for efficient X-ray detection.

Other attempts to improve the sensitivity and limit the dark current of semiconductor MHP detectors have focused on improved control over solution synthesis.

Similar synthetic approaches have also been utilized for the fabrication of devices based on large-sized semiconductors with 0D structures, such as $MA_3Bi_2I_9$ (ref.

Utilizing this technique, an MHP-based digital flat-panel detector was fabricated on a thin-film transistor substrate to function as a highly sensitive X-ray imager.

Although the dark currents for these devices were still fairly high, the ability to effectively form single crystals between the electrodes should allow this property to be improved by more careful and controlled fabrication.

Perovskite semiconductor detectors in single-photon mode

Solution-grown α-$FAPbI_3$ detectors have already demonstrated the first spectroscopic-grade response of γ-rays in the single-photon mode [14].

Solution-grown $CsPbBr_3$ single-crystal detectors demonstrated a similar spectroscopic response at low temperatures (−53 °C) [15].

After improvements in interfacial and surface conditions, a similar Cl-alloyed compound ($MAPbBr_{2.85}Cl_{0.15}$) demonstrated a spectroscopic α-particle response in the single-photon mode [16, 17].

With holes, excessive electron trapping was identified when the detector operated in electron-collecting conditions [18], with $\mu\tau_e$ being approximately half of $\mu\tau_h$ (ref.

In addition to Pb-based MHPs, perovskite-relevant semiconductors based on other heavy metals (Hg and Tl) have demonstrated promising hard radiation detection performance in the single-photon detection mode, such as the defect antiperovskite $Hg_3Q_2X_2$ (Q = S, Se, Te; X = Cl, Br, I) [19] and 2D antiperovskites $TlSn_2I_5$ (ref.

These Hg- and Tl-based semiconductors showed reasonable electron mobilities (reaching around 100 $cm^2 V^{-1} s^{-1}$) and demonstrated γ-ray responses distinguishable from the background.

Outlook

MHP semiconductors are the only materials at present that promise to reform conventional X- and γ-ray detectors with lower cost and higher performance.

Depending on the two detection schemes for X- and γ-ray detection, the optimization of the material properties should fit specific application criteria.

These standards are essential to obtain a reliable detector characterization for commercial applications; however, strictly following these procedures might not be practical as a routine approach in research laboratories.

A standard protocol for measuring and reporting X-ray detectors that is suitable for research laboratories is required.

We also note that the remarkably high X-ray sensitivity in halide perovskites is mostly attributed to the poorly understood large photoconductive gain effects.

The device stability of MHP detectors under high-dose radiation and high electric field must be clarified, as it is particularly important for commercialization.

Self-wavelength Shifting in Two-Dimensional Perovskite for Sensitive and Fast Gamma-Ray Detection

DOI: https://doi.org/10.1038/s41467-023-38545-y

This article is licensed under a Creative Commons Attribution 4.0 International License, which permits use, sharing, adaptation, distribution and reproduction in any medium or format, as long as you give appropriate credit to the original author(s) and the source, provide a link to the Creative Commons license, and indicate if changes were made. The images or other third party material in this article are included in the article's Creative Commons license, unless indicated otherwise in a credit line to the material. If material is not included in the article's Creative Commons license and your intended use is not permitted by statutory regulation or exceeds the permitted use, you will need to obtain permission directly from the copyright holder. To view a copy of this license, visit http://creativecommons.org/licenses/by/4.0/.

Originators: Jin, T., Liu, Z., Luo, J. *et al.* Self-wavelength shifting in two-dimensional perovskite for sensitive and fast gamma-ray detection. *Nat Commun* 14, 2808 (2023). https://doi.org/10.1038/s41467-023-38545-y

Abstract-Summary

Lead halide perovskites have recently emerged as promising X/γ-ray scintillators.

The small Stokes shift of exciton luminescence in perovskite scintillators creates problems for the light extraction efficiency and severely impedes their applications in hard X/γ-ray detection.

We demonstrate the intrinsic strain in 2D perovskite crystals as a general phenomenon, which could be utilized as self-wavelength shifting to reduce the self-absorption effect without sacrificing the radiation response speed.

This work provides a new paradigm for suppressing the self-absorption effect in scintillators and may facilitate the application of perovskite scintillators in practical hard X/γ-ray detections.

Introduction

The recently reported lead halide perovskite scintillators have demonstrated excellent properties for soft X-ray detection (<70 keV) because of the convenient solution synthesis process, considerable stopping power, low detection limit, and short radioluminescence (RL) lifetime [20–22].

For hard X-rays and γ-rays, the corresponding linear attenuation coefficients for scintillators are much smaller than that of soft X-rays, and the typical thickness of lead halide perovskite scintillators (4–120 μm) is obviously inadequate.

Although the self-absorption could be suppressed, the RL lifetime is severely extended from hundreds of nanoseconds to milliseconds, which strongly impedes their deployment in CT imaging and time-of-flight γ-ray detections.

We posit the intrinsic strain in 2D perovskite crystals as a general phenomenon, which could be utilized to suppress the self-absorption effect without sacrificing the luminescence lifetime.

We successfully applied the SWS scintillators in high energy γ-ray detection and demonstrated the imaging reconstruction in PET, which is the first time ever reported for perovskite scintillators.

Results

To further confirm the structure difference between the surface and the interior of the 2D perovskites, we performed GIWAXS analysis and grazing-incidence X-ray diffraction (GIXRD) measurement with various incident angles.

The lattice structure information at different depths on the surface of the 2D perovskite single crystal can be obtained by varying the X-ray incident angle.

The γ-ray light yields of PEA_2PbBr_4 and BA_2PbBr_4 are calculated to be 38,800 photons/MeV and 17,400 photons/MeV. The light yield of PEA_2PbBr_4 is also much higher than previously reported perovskite scintillators ($CsPbBr_3$ nanosheets, 21,000 photons/MeV [21]) and other composite-based scintillators with FRET mechanisms (gadolinium oxide–polymer nanocomposites, 27,000 photons/MeV [23]).

The thickness of the current 2D perovskite crystal is still not enough to reach high detection efficiency of 511 keV photons.

Another improvement is that the 2D perovskite sheet crystals could be aligned vertically toward the incident photons, i.e. the sheet plane is parallel to the incident direction of γ rays.

Discussion

We posit the intrinsic strain in 2D perovskite crystals as a general phenomenon, which could be utilized to decrease the self-absorption effect without sacrificing the RL lifetime.

The exemplary optical transparency and fast response speed enable their application in high-energy γ-ray detection.

On the practical side, we successfully demonstrated the first imaging reconstruction of source using a pair of 2D perovskite crystals as detectors in coincidence.

This work provides a new paradigm for suppressing the self-absorption effect in perovskite scintillators and bridges the perovskite scintillators and practical hard X-ray as well as γ-ray detections.

Methods

PEA_2PbBr_4 SCs XRD measurements were performed using a Philips diffractometer (X pert pro-MRD) with a step of 0.017° and step time of 10.16 s. The lines used were Cu $K\alpha_1$ and Cu $K\alpha_2$ with wavelengths of 1.54060 Å and 1.54443 Å, respectively.

GIXRD measurements were performed using a PANalytical B.V. diffractometer (X Pert Powder) with a step of 0.0052° and step time of 1.016 s. The lines used were Cu $K\alpha_1$ and Cu $K\alpha_2$ with wavelengths of 1.54060 Å and 1.54443 Å, respectively.

All test samples and crystals were coupled to a pair of 4×4 mm^2 Si-PMs (HAMAMATSU S14160-4050HS) with optical glue (Dow corning RTV 3145) with single-ended readout.

A pair of 2D perovskite crystals worked in the coincidence mode, which ensured that the collected data were generated by the solution.

Colour Selective Control of Terahertz Radiation Using Two-Dimensional Hybrid Organic Inorganic Lead-Trihalide Perovskites

DOI: https://doi.org/10.1038/s41467-017-01517-0

This article is licensed under a Creative Commons Attribution 4.0 International License, which permits use, sharing, adaptation, distribution and reproduction in any medium or format, as long as you give appropriate credit to the original author(s) and the source, provide a link to the Creative Commons license, and indicate if changes were made. The images or other third party material in this article are included in the article's Creative Commons license, unless indicated otherwise in a credit line to the material. If material is not included in the article's Creative Commons license and your intended use is not permitted by statutory regulation or exceeds the permitted use, you will need to obtain permission directly from the copyright holder. To view a copy of this license, visit http://creativecommons.org/licenses/by/4.0/.

Originators: Chanana, A., Zhai, Y., Baniya, S. *et al.* Colour selective control of terahertz radiation using two-dimensional hybrid organic inorganic lead-trihalide perovskites. *Nat Commun* 8, 1328 (2017). https://doi.org/10.1038/s41467-017-01517-0

Abstract-Summary

We demonstrate an innovative approach for controlling the propagation properties of terahertz (THz) radiation, through use of both the excitation optical wavelength (colour) and intensity.

The absorption properties of these materials in the visible range can be tuned by changing the number of inorganic atomic layers in between the organic cation layers.

Optical absorption in 2D perovskites occurs over a broad spectral range above the bandgap, resulting in free carrier generation, as well as over a narrow spectral range near the bandedge due to exciton formation.

By patterning multiple 2D perovskites with different optical absorption properties onto a single device, we demonstrate both colour selective modulation and focusing of THz radiation.

Introduction

All-optical approaches are particularly attractive because they offer the potential for high-speed modulation and switching, and can be fabricated straightforwardly on conventional semiconductor substrates.

In such devices, modulation of THz radiation can be realized when the incident optical control beam has a photon energy above the bandgap, where the generation of charge carriers leads to increased absorption of the incident THz radiation.

High peak fluence optical pulses are needed to generate sufficient photocarriers for modulating the THz response.

Compared to three-dimensional (3D) $CH_3NH_3PbI_3$ perovskites, two-dimensional (2D) perovskites self-assemble into alternating organic and inorganic layers that form natural 'multiple-quantum wells', with highly efficient photocurrent [24] and tunable optoelectronic characteristics [25].

We exploit the properties of 2D hybrid lead-trihalide perovskites – $(C_6H_5C_2H_4NH_3)_2PbI_4$(PEPI) – and its Ruddlesden–Popper (R–P) mixtures with $CH_3NH_3PbI_3$ [26], to enable colour selective THz modulation using plasmonic structures fabricated on a semiconductor substrate.

Results

We initially measured the transmitted broadband THz radiation through different samples, while optically exciting the perovskite side of the sample with filtered radiation from a halogen lamp (360–800 nm).

It is the interface between the silicon and perovskite that dramatically enhances broadband THz absorption upon optical illumination.

To better understand the excitation wavelength dependence on the THz absorption in the 2D perovskites/Si samples, we optically illuminated each sample using a series of 50 nm wide bandpass filters ranging from 450 to 800 nm.

THz absorption in these samples increases appreciably only when the wavelength of the incident optical radiation corresponds to the wavelength of the exciton resonance, but not at higher energies that corresponds to band-to-band transition.

We note that excitons do not dominate the absorption in 3D $MAPbI_3$ perovskites and, thus, we do not observe an analogous wavelength (colour) selective control of the THz absorption.

Discussion

We have demonstrated that 2D hybrid organic–inorganic lead-trihalide perovskites offer a number of unique characteristics for fabricating active THz devices that distinguish them from structures fabricated using conventional semiconductors.

Using a plasmonic structure over which multiple perovskites are deposited, we have shown near complete suppression of transmission resonances using only relatively low intensity narrow-band excitation from a halogen lamp.

This approach overcomes the limitations encountered with existing modulators: the devices require only relatively simple fabrication techniques [27–36] and when compared to optically induced modulation, THz absorption now depends upon the excitation optical wavelength and a halogen lamp can be used instead of an ultrafast laser source.

Methods

Fluorescence spectra were taken with an optical fiber based spectrometer (Ocean Optics USB4000ES) by exciting the films on glass substrates with a 447 nm semiconductor laser (50 mW).

The 2D perovskite films were deposited on Si and KBr substrates and placed in a cryostat in which the temperature could be varied from 45 to 300 K. The pump and probe beams were co-incident on the sample films, and the transmission, T, and change of the transmitted probe beam (ΔT) was detected by a DLaTGS detector.

A thin layer of PDMS mask was then attached onto the silicon wafer before spin-coating the perovskite precursor on top.

The wafer to remove the residual solvent, the uncovered part was uniformly coated with perovskite material and the PDMS mask was removed.

A second off-axis parabolic mirror placed after the filter was used to focus the transmitted THz radiation onto the detection crystal for measurement via electro-optic sampling.

Electron Tunneling at the Molecularly Thin 2D Perovskite and Graphene van der Waals Interface

DOI: https://doi.org/10.1038/s41467-020-19331-6

This article is licensed under a Creative Commons Attribution 4.0 International License, which permits use, sharing, adaptation, distribution and reproduction in any medium or format, as long as you give appropriate credit to the original author(s) and the source, provide a link to the Creative Commons license, and indicate if changes were made. The images or other third party material in this article are included in the article's Creative Commons license, unless indicated otherwise in a credit line to the material. If material is not included in the article's Creative Commons license and your intended use is not permitted by statutory regulation or exceeds the permitted use, you will need to obtain permission directly from the copyright holder. To view a copy of this license, visit http://creativecommons.org/licenses/by/4.0/.

Originators: Leng, K., Wang, L., Shao, Y. *et al.* Electron tunneling at the molecularly thin 2D perovskite and graphene van der Waals interface. *Nat Commun* 11, 5483 (2020). https://doi.org/10.1038/s41467-020-19331-6

Abstract-Summary

A major bottleneck to device performance is the high charge injection barrier caused by organic molecular layers on its basal plane, thus the best performing device currently relies on edge contact.

By leveraging on van der Waals coupling and energy level matching between two-dimensional Ruddlesden-Popper perovskite and graphene, we show that the plane-contacted perovskite and graphene interface presents a lower barrier than gold for charge injection.

Field effect transistors fabricated on molecularly thin Ruddlesden-Popper perovskite using graphene contact exhibit electron mobilities ranging from 0.1 to 0.018 $cm^2V^{-1}s^{-1}$ between 1.7 to 200 K. Scanning tunneling spectroscopy studies reveal layer-dependent tunneling barrier and domain size on few-layered Ruddlesden-Popper perovskite.

Introduction

Identifying an electrode material with the interface energy alignment to favor low resistance contact on 2D perovskites is needed for the fabrication of FET [37].

Little is known about this hybrid interface of 2D perovskite and graphene including their band alignment, the surface structure of 2D perovskites on graphene, and the dynamics of interfacial charge transfer.

We investigate charge injection and electrical transport in molecularly thin single crystalline 2D $(C_4H_9NH_3)_2(CH_3NH_3)_3Pb_4I_{13}$ ($(BA)_2(MA)_3Pb_4I_{13}$, $n = 4$) RPP using van der Waals-contacted graphene (G) as the electrode in FET devices.

The atomic structure of the 2D perovskites on G and Au was examined using scanning tunneling microscope (STM), where a higher degree of substrate-induced disorder was found on Au than on G. Domain structure on molecularly thin 2D perovskites was observed to be thickness-dependent, and layer-dependent tunneling gaps were investigated using scanning tunneling spectroscopy (STS).

Results

Establishing the energy alignment between RPP and the contact electrodes (Au or G) provides the basic framework for understanding contact barriers and charge transfer behavior across the interface.

A higher degree of surface crystalline order was observed for RPP on G compared to Au, which is attributed to the high-quality van der Waals interface formed between RPP and G. Next, the electronic structure of RPP flake from monolayer to bulk thickness is investigated.

For RPP flake of similar thickness, the energy gap is measured to be larger on Au than on G. Taking into account the higher quality domain order and smaller tunneling gap for perovskites supported on G than on Au, perovskite flakes with few-layer thickness (<10 nm), supported on G, were selected for device fabrication.

Besides the energy level matching between graphene's Fermi level with the conduction bands in RPP, the ultrafast charge transfer may be enabled by photo-assisted quantum tunneling across the RPP/G interface.

Discussion

Our studies show that graphene serves as a low barrier and gate-tunable van der Waals contact for $n = 4$ RPP where we observed the classic tunneling mechanism of field emission at high bias voltage and direct tunneling at low bias.

A considerably lower barrier for charge injection was observed on RPP/G as compared to the RPP/Au interface.

Highly ordered domains were observed under the STM for ultrathin RPP on graphene, where the thickness-dependent tunneling gap was observed.

The presence of an ordered array of organic chains interleaving perovskite crystals hints at an interesting, yet unexplored possibility of using molecularly thin RPP both as both a tunneling barrier and spin generator on graphene.

Methods

Two methods based on $n = 4$ RPP exfoliated on different substrates (pure Si and G) were carried out to calculate work function of $n = 4$ RPP.

This allows us to evaluate the work function difference between $n = 4$ RPP and G by KPFM measurement (Bruker Dimension Icon).

Different from method one, here the external voltage was applied to the sample with the tip grounded.

A positive KPFM signal indicates a higher work function of the sample relative to the probe.

By the same calculation method, the Au-contacted RPP device showed a Φ of 1.82 eV at 1.7 K. The ellipsometry studies were carried out on J. A. Woollam ellipsometer which has monochromatic radiation photon with energies ranging from 0.6 to 6.5 eV. The sample is measured in high vacuum cryostat with pressure of 1×10^{-9} Torr with incidence angles (θ) at 70°.

Perovskite Semiconductors for Room-Temperature Exciton-Polaritonics

DOI: https://doi.org/10.1038/s41563-021-01035-x

Originators: Su, R., Fieramosca, A., Zhang, Q. *et al.* Perovskite semiconductors for room-temperature exciton-polaritonics. *Nat. Mater.* 20, 1315–1324 (2021). https://doi.org/10.1038/s41563-021-01035-x

Abstract-Summary

In this Perspective, we discuss fundamental aspects of perovskite semiconductors for exciton-polaritons and review the recent rapid experimental advances using lead-halide perovskites for room-temperature polaritonics, including the experimental realization of strong light–matter interaction using various types of microcavities as well as reaching the polariton condensation regime in planar microcavities and lattices.

Main

Owing to their mixed light and matter nature, microcavity exciton-polaritons have inherited remarkable properties from both photons and excitons, such as low effective mass and strong nonlinearity [38, 39], underpinning their crucial roles not only in fundamental sciences, such as in the study of polariton condensation [40], superfluidity [41, 42], quantum vortices [43] and many more exotic phenomena, but also in technologically relevant photonic and optoelectronic devices, such as low threshold polariton lasers [44, 45], all-optical polariton logic devices [46, 47] and neuromorphic computing [48].

In the strong coupling regime, coherent energy exchange between light and matter becomes reversible, giving rise to two new hybrid states called lower and upper exciton-polaritons with a typical anticrossing feature in the dispersion.

Polariton lasing that is like Bose–Einstein condensation operates well below the Mott density $n < n_M$, and Bardeen–Cooper–Schrieffer-like polariton lasing approaches or even exceeds the Mott density $n \geq n_M$ (ref [49]), wherewhere k_B is the Boltzmann constant, T is the temperature, a_B is the exciton Bohr radius and E_b is the exciton binding energy.

The second threshold behaviour is usually difficult to observe in single-mode exciton-polariton systems with large Rabi splitting where the energy spacing between the ground state of lower exciton-polaritons and pure cavity modes is quite large.

Fundamental aspects of perovskite semiconductors for exciton-polaritons

Although these requirements are necessary conditions, if not accompanied by strong optical nonlinearities and high uniformity, it is not realistic to fully exploit the hybrid light–matter nature of exciton-polaritons [39], and high-quality single crystals of both 2D Ruddlesden–Popper [50–53] and 3D bulk chloride and bromide perovskites [54–56] turned out to be excellent candidates.

The hybrid 2D perovskite, benefiting from the quantum and dielectric confinement owing to the natural QW structure, shows an exciton binding energy up to 470 meV (ref [51]).

Wannier–Mott-type excitons with large Bohr radius, which are generally expected for perovskites, are highly sought for realistic room-temperature polaritonic applications.

The bulk 3D ones with lower binding energies benefit from a bigger exciton spatial extension compared with the 2D perovskite, which instead is limited by the large oscillator strength and binding energy.

Since the research on strong light–matter interactions in perovskites is still in its nascent stage, modified perovskite materials showing suitable oscillator strength, high crystalline quality and highly interacting Wannier–Mott excitons could finally satisfy the stringent requirements imposed by room-temperature operation.

Exciton-polaritons in perovskite photonic structures

Benefiting from the recent tremendous progress in photovoltaic device engineering, the crystallinity of 2D hybrid organic–inorganic perovskites has been significantly improved by new synthesis methods, such as an antisolvent vapour crystallization [52], which allowed the observation of the strong coupling regime in double DBR planar microcavities, reaching higher Q of over 2,000 (ref [57]).

In the quest for coherent ground-state condensation in a 2D perovskite at elevated temperatures, the rational control of the polariton dispersion with the aim of controlling the polariton relaxation dynamics could represent a suitable alternative approach.

A 2D PEPI film was directly patterned into photonic crystal structure to make perovskite-based subwavelength metasurfaces [58], showing the possibility of tailoring the polaritonic dispersion at room temperature.

The 2D perovskites exhibit disadvantages in transport properties rooted in the organic ligands that could prevent future electrically pumped polaritonic devices.

Strong coupling between excitons and surface plasmons in perovskites was first demonstrated by spin-coating a 2D PEPI perovskite film onto a silver layer with a thin SiO_2 spacer [59], while similar results were demonstrated by depositing the same material on a silver grating [60].

Polariton condensates in perovskite microcavities

Although strong light–matter interaction has been well documented in lead-halide perovskites, polariton condensation has always been a challenge among many optical gain materials operating at room temperature, including organics, hybrids and perovskites microcavities.

In 2017, the first demonstration of non-equilibrium polariton condensation in lead-halide perovskites was realized in a new type of all-inorganic $CsPbCl_3$ perovskite planar microcavity at room temperature [61].

Along with the superior perovskite crystal quality and microcavity quality, which ensure long polariton lifetimes for relaxation, these advantages promote the realization of non-equilibrium polariton condensation in all-inorganic perovskite microcavities even at room temperature.

Non-equilibrium polariton condensation was also successfully achieved with 2D PEPI perovskites in a planar microcavity at 4 K (ref [62]).

These single-crystalline 2D perovskite layers possess much better crystalline quality; a flat surface upon exfoliation; robust excitonic properties, which provide a microcavity quality factor larger than 1,000; a longer polariton lifetime; and a higher damaging threshold, underpinning the realization of non-equilibrium polariton condensation.

Perspectives and outlook

Halide perovskites have been demonstrated to be remarkable platforms to bring the physics of exciton-polaritons to room temperature.

Thanks to their robust excitonic properties, rich material engineering approaches and their highly nonlinear response [63], these perovskite semiconductors appear to be ideal candidates for future development of polaritonic devices.

From the fundamental and practical viewpoints, the strong nonlinearity in perovskite systems, as the cornerstone of polaritonics, is of great importance and could lead, with proper enhancement, to exploring the possibility of bringing polaritonics into a quantum regime.

The strong spin-dependent nonlinearity in perovskite systems [63] could potentially introduce significant optical Zeeman splitting to break the time-reversal symmetry, facilitating the realization of topological protected exciton polaritons at room temperature, even without the need of a magnetic field.

Plasmonic–Perovskite Solar Cells, Light Emitters, and Sensors

DOI: https://doi.org/10.1038/s41378-021-00334-2

This article is licensed under a Creative Commons Attribution 4.0 International License, which permits use, sharing, adaptation, distribution and reproduction in any medium or format, as long as you give appropriate credit to the original author(s) and the source, provide a link to the Creative Commons license, and indicate if changes were made. The images or other third party material in this article are included in the article's Creative Commons license, unless indicated otherwise in a credit line to the material. If material is not included in the article's Creative Commons license and your intended use is not permitted by statutory regulation or exceeds the permitted use, you will need to obtain permission directly from the copyright holder. To view a copy of this license, visit http://creativecommons.org/licenses/by/4.0/.

Originators: Ai, B., Fan, Z. & Wong, Z.J. Plasmonic–perovskite solar cells, light emitters, and sensors. *Microsyst Nanoeng* 8, 5 (2022). https://doi.org/10.1038/s41378-021-00334-2

Abstract-Summary

In optoelectronics, perovskite materials are particularly attractive due to their excellent absorption, emission, and carrier transport properties, which lead to the improved performance of solar cells, light-emitting diodes (LEDs), lasers, photodetectors, and sensors.

When perovskite materials are coupled with plasmonic structures, the device performance significantly improves owing to strong near-field and far-field optical enhancements, as well as the plasmoelectric effect.

We review recent theoretical and experimental works on plasmonic perovskite solar cells, light emitters, and sensors.

This review also lays out challenges and future directions for the plasmonic perovskite research field toward next-generation optoelectronic technologies.

Introduction

The most prominent application of halide perovskites is as light-absorbing materials in solar cells.

Miyasaka and coworkers first applied halide perovskite materials in dye-sensitized solar cells in 2009 [64].

Halide perovskites have also been used as light-emitting materials in LEDs and lasers.

Benefiting from the strong light absorption of halide perovskites, new applications have been found in regard to sensing and signal detection [65–69].

The ability of plasmonics to concentrate light at the nanoscale and enhance light–matter interaction holds the key to significantly improving the performance of perovskite optoelectronic devices.

We review the recent works of plasmonic and perovskite-based solar cells, LEDs, lasers, sensors and other applications to provide a more comprehensive summary of plasmonic implementation and effects on halide perovskite devices.

The second section (Plasmonic–perovskite light emitters) is focused on exploiting the plasmonic nanostructures in perovskite light emitters, including enhanced spontaneous emission and lasers.

Plasmonic–perovskite solar cells

The optimum NP distance was ~300 nm; (b) the effect of NP randomness, angle of incidence, and polydispersity was very limited on the calculated enhancement; (c) the Al nanospheres with radii of 18 nm and a gap distance of 65 nm had the optimum performance; (d) the photocurrent was largely enhanced (up to 4.0% corresponding to photocurrent density (J_{ph}) = 21.22 mA/cm^2) when NPs were in the HTL (PEDOT:PSS, on top of the perovskite layer)and (e) the largest absorption enhancement was achieved when Al and Ag spheres were both used in the PEDOT:PSS layer and perovskite layer, respectively.

The following conclusions are extracted based on the analysis of previous works: (1) The best position for using plasmonic NPs in PSCs is the perovskite-m-TiO_2 interface because m-TiO_2 is one of the best ETL materials that has good conductivity and a large surface area, and the plasmonic effects can both enhance the light–matter interaction of the perovskite and m-TiO_2 layers. (2) NPs with core–shell structures, modified shapes, and dimers show stronger effects than simple spherical NPs.

Plasmonic–perovskite light emitters

Plasmonic NPs and films with Au and Ag materials have been used in perovskite LEDs and have achieved impressive PL and EQE enhancements.

The properties of plasmonic-based perovskite lasers, such as the threshold, wavelength, polarization, and laser emission characteristics, can be facilely controlled by the rational design of the coupling between perovskites and plasmonic nanostructures.

Plasmonic lasers with a low mode volume have been realized by coupling the perovskite with plasmonic NPs and films.

Wang and others demonstrated plasmonic-enhanced laser emissions from all-inorganic lead-halide perovskites [70].

The lasing threshold and stimulated emission of the Al-$MAPbBr_3$ hybrid system were dramatically reduced by more than 27% and improved by 10 times, respectively, due to the plasmonic effects of the Al NPs.

The Purcell effect can be enhanced by plasmonic resonances, which increases the light–matter interaction and emission intensity while reducing the emission lifetime, thus enabling faster lasers and faster LEDs.

Plasmonic–perovskite sensors

Plasmon-enhanced light harvesting was improved on the perovskite–graphene interface, which led to a higher PD operation speed and carrier extraction efficiency than devices without Au NPs.

Ghosh and others reported an air-processed high-performance self-powered hybrid perovskite PD whose ETL was embedded with plasmonic Ag NPs [71].

The hybrid plasmonic–perovskite nanostructures showed a 43-fold increase in the photocurrent, which was due to the enhanced E-fields around the Au NPs.

The responsivity, noise equivalent power (NEP), and response speed of the perovskite with Au NPs were ~2063 A/W, ~1.93×10^{-14} W/$Hz^{1/2}$, and ~300 ns, respectively, which were approximately 4.1-, 15.6-, and 2.8-fold increases, respectively, compared with those of a perovskite PD without NPs.

By integrating plasmonic nanostructures with perovskites, enhanced light absorption can be attained, which opens the door for high-performance photodetectors.

Other plasmonic–perovskite applications

As the Ag content is increased, the UV-induced photoreactivity first increases, reaching a maximum at a Ag content of ~1.7 wt%, and then decreases.

Visible light-induced photoreactivity shows a monotonic positive correlation with Ag contents from 0.4 wt% to 2.8 wt%.

That energy migrated from the perovskite NCs toward the Ag NCs.

Zhou and others reported a polarization-dependent modulation based on hybrid perovskite plasmon-induced transparency (PIT) [72].

Conclusions and outlook

Researchers have demonstrated the use of plasmonics to enhance the performance of perovskite-based solar cells, light emitters, photodetectors, sensors, and many other applications.

We observed several trends from the vast amount of literature on NP-assisted PSCs that we examined: (1) the PCE of PSCs can be significantly enhanced by plasmonic NPs; (2) PSCs with plasmonic NPs incorporated at the perovskite-m-TiO_2 interface or simultaneously at multiple positions can potentially achieve the largest PCE enhancement; (3) NPs with a core–shell structure, shape modification, and dimer show stronger effects than that of simple spherical NPs; (4) alloyed NPs show stronger enhancement than that of a single material; and (5) the size and concentration have different effects on the current, voltage, and FF; thus, they need to be optimized carefully.

To increasing light absorption and photocurrent generation, plasmonic NPs and films can also enhance the light emission of perovskites, which benefits the applications of LEDs and lasers.

Bibliography

1. Protesescu, L., et al.: Nanocrystals of cesium lead halide perovskites ($CsPbX_3$, X = Cl, Br, and I): novel optoelectronic materials showing bright emission with wide color gamut. Nano Lett. **15**, 3692–3696 (2015)
2. Chen, X., et al.: Centimeter-sized Cs_4PbBr_6 crystals with embedded $CsPbBr_3$ nanocrystals showing superior photoluminescence: nonstoichiometry induced transformation and light-emitting applications. Adv. Funct. Mater. **28**, 1706567 (2018)
3. Li, J., et al.: Hexagonal boron nitride epitaxial layers as neutron detector materials. Nucl. Inst. Methods Phys. Res. A **654**, 417–420 (2011)
4. Andričević, P., et al.: Kilogram-scale crystallogenesis of halide perovskites for gamma-rays dose rate measurements. Adv. Sci. **8**(2), 2001882 (2021)
5. Webster, J.R.P., et al.: Reflectometry techniques on the second target station at ISIS: methods and science. Eur. Phys. J. Plus **126**(11), 112 (2011)
6. Huang, H., Abbaszadeh, S.: Recent developments of amorphous selenium-based X-ray detectors: a review. IEEE Sens. J. **20**, 1694–1704 (2020)
7. Chen, Z., Zhu, Y., He, Z.: Intrinsic photopeak efficiency measurement and simulation for pixelated CdZnTe detector. Nucl. Instr. Meth. Phys. Res. A **980**, 164501 (2020)
8. Wu, H., Ge, Y., Niu, G., Tang, J.: Metal halide perovskites for X-ray detection and imaging. Matter **4**, 144–163 (2021)
9. Pan, W., et al.: $Cs_2AgBiBr_6$ single-crystal X-ray detectors with a low detection limit. Nat. Photon. **11**, 726–732 (2017)
10. Yin, L., et al.: Controlled cooling for synthesis of $Cs_2AgBiBr_6$ single crystals and its application for X-ray detection. Adv. Opt. Mater. **7**, 1900491 (2019)
11. Yang, B., et al.: Heteroepitaxial passivation of $Cs_2AgBiBr_6$ wafers with suppressed ionic migration for X-ray imaging. Nat. Commun. **10**, 1989 (2019)
12. Keshavarz, M., et al.: Tuning the structural and optoelectronic properties of $Cs_2AgBiBr_6$ double-perovskite single crystals through alkali-metal substitution. Adv. Mater. **32**, 2001878 (2020)
13. Stranks, S.D., Snaith, H.J.: Metal-halide perovskites for photovoltaic and light-emitting devices. Nat. Nanotechnol. **10**, 391–402 (2015)

14. Yakunin, S., et al.: Detection of gamma photons using solution-grown single crystals of hybrid lead halide perovskites. Nat. Photon. **10**, 585–589 (2016)
15. Dirin, D.N., Cherniukh, I., Yakunin, S., Shynkarenko, Y., Kovalenko, M.V.: Solution-grown $CsPbBr_3$ perovskite single crystals for photon detection. Chem. Mater. **28**, 8470–8474 (2016)
16. Tan, R., et al.: Improved radiation sensing with methylammonium lead tribromide perovskite semiconductors. Nucl. Instr. Meth. Phys. Res. A **986**, 164710 (2021)
17. Xu, Q., et al.: Detection of charged particles with a methylammonium lead tribromide perovskite single crystal. Nucl. Instr. Meth. Phys. Res. A **848**, 106–108 (2017)
18. He, Y., et al.: $CsPbBr_3$ perovskite detectors with 1.4% energy resolution for high-energy γ-rays. Nat. Photon. **15**, 36–42 (2021)
19. He, Y., et al.: Defect antiperovskite compounds $Hg_3Q_2I_2$ ($Q = S$, Se, and Te) for room-temperature hard radiation detection. J. Am. Chem. Soc. **139**, 7939–7951 (2017)
20. Chen, Q., et al.: All-inorganic perovskite nanocrystal scintillators. Nature **561**, 88–93 (2018)
21. Zhang, Y., et al.: Metal halide perovskite nanosheet for X-ray high-resolution scintillation imaging screens. ACS Nano **13**, 2520–2525 (2019)
22. Cao, F., et al.: Shining emitter in a stable host: design of halide perovskite scintillators for X-ray imaging from commercial concept. ACS Nano. **14**, 5183–5193 (2020)
23. Cai, W., et al.: Synthesis of bulk-size transparent gadolinium oxide-polymer nanocomposites for gamma ray spectroscopy. J. Mater. Chem. C. **1**, 1970–1976 (2013)
24. Klots, A.R., et al.: Probing excitonic states in suspended two-dimensional semiconductors by photocurrent spectroscopy. Sci. Rep. **4**, 6608 (2014)
25. Mitzi, D.B., Feild, C.A., Harrison, W.T.A., Guloy, A.M.: Conducting tin halides with a layered organic-based perovskite structure. Nature **369**, 467–469 (1994)
26. Stoumpos, C.C., et al.: Ruddlesden-Popper hybrid lead iodide perovskite 2D homologous semiconductors. Chem. Mater. **28**, 2852–2867 (2016)
27. Tao, H., et al.: Reconfigurable terahertz metamaterials. Phys. Rev. Lett. **103**, 147401 (2009)
28. Zhu, W.M., et al.: Microelectromechanical maltese-cross metamaterial with tunable terahertz anisotropy. Nat. Commun. **3**, 1274 (2012)
29. Pitchappa, P., et al.: Micro-electro-mechanically tunable metamaterial with enhanced electro-optic performance. Appl. Phys. Lett. **104**, 151104 (2014)
30. Unlu, M., et al.: Switchable scattering meta-surfaces for broadband terahertz modulation. Sci. Rep. **4**, 5708 (2014)
31. Kleine-Ostmann, T., Dawson, P., Pierz, K., Hein, G., Koch, M.: Room-temperature operation of an electrically driven terahertz modulator. Appl. Phys. Lett. **84**, 3555–3557 (2004)
32. Chen, H.-T., et al.: Active terahertz metamaterial devices. Nature **444**, 597–600 (2006)
33. Chen, H.-T., et al.: A metamaterial solid-state terahertz phase modulator. Nat. Photon. **3**, 148–151 (2009)
34. Ju, L., et al.: Graphene plasmonics for tunable terahertz metamaterials. Nat. Nanotechnol. **6**, 630–634 (2011)
35. Lee, S.H., et al.: Switching terahertz waves with gate-controlled active graphene metamaterials. Nat. Mater. **11**, 936–941 (2012)
36. Sensale-Rodriguez, B., et al.: Broadband graphene terahertz modulators enabled by intraband transitions. Nat. Commun. **3**, 1787 (2012)
37. Wang, S., Sakurai, T., Wen, W., Qi, Y.: Energy level alignment at interfaces in metal halide perovskite solar cells. Adv. Mater. Inter. **5**, 1800260 (2018)
38. Deng, H., Haug, H., Yamamoto, Y.: Exciton-polariton Bose-Einstein condensation. Rev. Mod. Phys. **82**, 1489–1537 (2010)
39. Sanvitto, D., Kéna-Cohen, S.: The road towards polaritonic devices. Nat. Mater. **15**, 1061–1073 (2016)
40. Kasprzak, J., et al.: Bose-Einstein condensation of exciton polaritons. Nature **443**, 409–414 (2006)
41. Amo, A., et al.: Superfluidity of polaritons in semiconductor microcavities. Nat. Phys. **5**, 805–810 (2009)

42. Lerario, G., et al.: Room-temperature superfluidity in a polariton condensate. Nat. Phys. **13**, 837–841 (2017)
43. Lagoudakis, K.G., et al.: Quantized vortices in an exciton–polariton condensate. Nat. Phys. **4**, 706–710 (2008)
44. Deng, H., Weihs, G., Snoke, D., Bloch, J., Yamamoto, Y.: Polariton lasing vs. photon lasing in a semiconductor microcavity. Proc. Natl Acad. Sci. USA **100**, 15318–15323 (2003)
45. Schneider, C., et al.: An electrically pumped polariton laser. Nature **497**, 348–352 (2013)
46. Ballarini, D., et al.: All-optical polariton transistor. Nat. Commun. **4**, 1778 (2013)
47. Zasedatelev, A.V., et al.: A room-temperature organic polariton transistor. Nat. Photon. **13**, 378–383 (2019)
48. Ballarini, D., et al.: Polaritonic neuromorphic computing outperforms linear classifiers. Nano Lett. **20**, 3506–3512 (2020)
49. Hu, J., et al.: Polariton laser in the Bardeen-Cooper-Schrieffer regime. Phys. Rev. X **11**, 011018 (2021)
50. Blancon, J.-C., et al.: Extremely efficient internal exciton dissociation through edge states in layered 2D perovskites. Science **355**, 1288–1292 (2017)
51. Blancon, J.C., et al.: Scaling law for excitons in 2D perovskite quantum wells. Nat. Commun. **9**, 2254 (2018)
52. Fieramosca, A., et al.: Tunable out-of-plane excitons in 2D single-crystal perovskites. ACS Photon. **5**, 4179–4185 (2018)
53. Fu, Y., et al.: Metal halide perovskite nanostructures for optoelectronic applications and the study of physical properties. Nat. Rev. Mater. **4**, 169–188 (2019)
54. Zhang, Q., Su, R., Liu, X., Xing, J., Sum, T.C., Xiong, Q.: High-quality whispering-gallery-mode lasing from cesium lead halide perovskite nanoplatelets. Adv Funct Mater **26**(34), 6238–6245 (2016). https://doi.org/10.1002/adfm.201601690
55. Akkerman, Q.A., et al.: Solution synthesis approach to colloidal cesium lead halide perovskite nanoplatelets with monolayer-level thickness control. J. Am. Chem. Soc. **138**, 1010–1016 (2016)
56. Shamsi, J., et al.: Colloidal synthesis of quantum confined single crystal $CsPbBr_3$ nanosheets with lateral size control up to the micrometer range. J. Am. Chem. Soc. **138**, 7240–7243 (2016)
57. Wang, J., Su, R., Xing, J., Bao, D., Diederichs, C., Liu, S., Liew, T.C.H., Chen, Z., Xiong, Q.: Room temperature coherently coupled exciton-polaritons in two-dimensional organic-inorganic perovskite. ACS Nano **12**(8), 8382–8389 (2018). https://doi.org/10.1021/acsnano.8b03737
58. Dang, N.H.M., et al.: Tailoring dispersion of room-temperature exciton-polaritons with perovskite-based subwavelength metasurfaces. Nano Lett. **20**, 2113–2119 (2020)
59. Symonds, C., et al.: Emission of hybrid organic-inorganic exciton/plasmon mixed states. Appl. Phys. Lett. **90**, 091107 (2007)
60. Niu, W., et al.: Image excitons and plasmon-exciton strong coupling in two-dimensional perovskite semiconductors. Phys. Rev. B **91**, 161303 (2015)
61. Su, R., et al.: Room-temperature polariton lasing in all-inorganic perovskite nanoplatelets. Nano Lett. **17**, 3982–3988 (2017)
62. Polimeno, L., et al.: Observation of two thresholds leading to polariton condensation in 2D hybrid perovskites. Adv. Opt. Mater. **8**, 2000176 (2020)
63. Fieramosca, A., et al.: Two-dimensional hybrid perovskites sustaining strong polariton interactions at room temperature. Sci. Adv. **5**, eaav9967 (2019)
64. Kojima, A., Teshima, K., Shirai, Y., Miyasaka, T.: Organometal halide perovskites as visible-light sensitizers for photovoltaic cells. J. Am. Chem. Soc. **131**, 6050–6051 (2009)
65. Deng, H., et al.: Flexible and semitransparent organolead triiodide perovskite network photodetector arrays with high stability. Nano Lett. **15**, 7963–7969 (2015)
66. Chen, K., Tüysüz, H.: Morphology-controlled synthesis of organometal halide perovskite inverse opals. Angew. Chem. Int. Ed. **54**, 13806–13810 (2015)
67. Zhuo, S., Zhang, J., Shi, Y., Huang, Y., Zhang, B.: Self-template-directed synthesis of porous perovskite nanowires at room temperature for high-performance visible-light photodetectors. Angew. Chem. Int. Ed. **54**, 5693–5696 (2015)

68. Miao, J., Zhang, F.: Recent progress on highly sensitive perovskite photodetectors. J. Mater. Chem. C **7**, 1741–1791 (2019)
69. Yu, X., Tsao, H.N., Zhang, Z., Gao, P.: Miscellaneous and perspicacious: hybrid halide perovskite materials based photodetectors and sensors. Adv. Opt. Mater. **8**, 2001095 (2020)
70. Wang, S., et al.: Maskless fabrication of aluminum nanoparticles for plasmonic enhancement of lead halide perovskite lasers. Adv. Opt. Mater. **5**, 1700529 (2017)
71. Ghosh, J., Natu, G., Giri, P.K.: Plasmonic hole-transport-layer enabled self-powered hybrid perovskite photodetector using a modified perovskite deposition method in ambient air. Org. Electron. **71**, 175–184 (2019)
72. Zhou, J.H., et al.: Ultrasensitive polarization-dependent terahertz modulation in hybrid perovskites plasmon-induced transparency devices. Photon. Res. **7**, 994–1002 (2019)

SPRINGER NATURE

GPSR Compliance

The European Union's (EU) General Product Safety Regulation (GPSR) is a set of rules that requires consumer products to be safe and our obligations to ensure this.

If you have any concerns about our products, you can contact us on ProductSafety@springernature.com

In case Publisher is established outside the EU, the EU authorized representative is:

Springer Nature Customer Service Center GmbH
Europaplatz 3
69115 Heidelberg, Germany

The manufacturer's authorised representative in the EU is Springer Nature Customer Service Centre GmbH, Europaplatz 3, 69115 Heidelberg, Germany. If you have any concerns regarding our products, please contact ProductSafety@springernature.com

Printed and bound by CPI Group (UK) Ltd, Croydon, CR0 4YY
01/10/2026
02225932-0001